Voyages in Conceptual Chemistry

Voyages in Conceptual Chemistry

Dan H. Barouch
HARVARD UNIVERSITY

With Preface by
Dudley Herschbach

Jones and Bartlett Publishers
Sudbury, Massachusetts

Boston   London   Singapore

Editorial, Sales, and Customer Service Offices
40 Tall Pine Drive
Sudbury, MA01776
508-443-5000
800-832-0034
info@jbpub.com
http://www.jbpub.com

Jones and Bartlett Publishers International
Barb House, Barb Mews
London W6 7PA
UK

Library of Congress Cataloging-in-Publication Data

   (not available at press time)

   ISBN: 0-7637-0308-7

Printed in the United States of America
00 99 98 97 96   10 9 8 7 6 5 4 3 2 1

to my family

Contents

Chapter 7

Quantum Theory, Atomic Structure, and Periodic Trends 121

Chapter 8
Bonding and Molecular Structure   141

Chapter 9
Special Topics   163

Introduction

Welcome to the fascination, the excitement, and the challenge of the world of chemistry. You are embarking on a creative voyage, an adventure, an exploration that will challenge your mind and change your perception of the world around you.

As students, please do not just memorize and recite, blindly learning formulas and performing numerous calculations. You must explore and think in order to blossom intellectually. By exploring you will discover, and through discoveries you will learn. Accordingly, this is not merely a problems book. This is not a compendium of recitation exercises, not a long list of numbers to plug into a long list of formulas, not a source of tedious homework sets. In none of these problems will you make numerical calculations. So PUT AWAY YOUR CALCULATORS!! Use these problems as springboards for conceptual thinking and qualitative understanding. You are a voyager in the world of chemistry, and this book is your tour guide.

These problems are designed to cover the principles taught in a typical first-year chemistry course. They ask you to solve puzzles, interpret observations, design experiments, and appreciate subtleties in the material. They also challenge you to utilize the versatility of general chemistry in applying your knowledge to solve problems in related subjects that you might study in the future, such as organic chemistry, quantum mechanics, biochemistry, physical chemistry, physiology, and medicine.

The type of conceptual thinking encouraged by this book is meant to supplement, not to replace, a regular textbook. I believe that the most effective chemistry education involves a balance of both quantitative and qualitative problems. Most current textbooks have numerous examples of the calculation-based questions but often lack qualitative "thought" problems. The American Chemical Society's Task Force for reforming the general chemistry curriculum has also recently concluded that conceptual exercises are essential for gaining a deep understanding of the material, and it has endorsed these *Voyages in Conceptual Chemistry* problems as examples of the types of problems that should be used in freshman courses.

The original edition of this problems book was designed in 1993 for use with the introductory chemistry course at Harvard. Based on our experience over the past three years, the book has been revised and expanded to include 150 problems covering all difficulty levels for the entire spectrum of topics. A difficulty level of A, B, or C is assigned to xiv

each problem. "A" problems are basic applications of usually one concept; "B" problems involve more difficult applications of more than one concept; and "C" problems either involve some subtleties or challenge you to synthesize the material in a more complex way. In all cases, "Hints" to the problems are located at the back of each chapter.

I would like to extend warm thanks and acknowledge the following individuals who were instrumental in the conception and creation of this book:

- **Dudley Herschbach**, founder of this project and proponent of conceptual understanding of chemistry, for his endless energy, enthusiasm, and vision;

- **Philip Anfinrud**, with whom I worked closely while I was teaching freshman chemistry, for his diligence and dedication to teaching;

- **James Davis**, with whom I have both studied and worked as a teaching fellow, for introducing me to the science of chemistry and the art of teaching, as well as for his boundless knowledge and always pleasant company;

- **Stephen Harrison** and **Andrew McMichael**, my undergraduate and graduate research advisors, who taught me both the principles and the practice of scientific research and thinking;

- **Fina Cañas**, my wonderful illustrator;

- **Paul Ma** and **Ashley Wivel**, my friends and consultants, for assistance in the production of the original edition of this book;

- **all the students** who have provided very useful feedback in the development of this book;

- **David Phanco** and the staff at Jones and Bartlett Publishers;

- and above all, **Eytan and Winifred Barouch**, my parents, for their love and support over the years.

Fasten your seatbelt, begin your exploration, and open your mind to the world of chemistry. Bon voyage.

Dan H. Barouch

Preface

Introductory courses in general chemistry and physics typically put much emphasis on solving numerical problems. Students certainly need to develop competence and confidence in solving such problems. But just as with other skillful arts, like music, dance, and sports, practice routines do not automatically produce happy results. Exercises overdone or sloppily done often induce dullness or downright bad habits. The usual textbook problems should bear a warning label: *Too much exposure to this stuff is dangerous to your mental health!*

The danger is manifested in at least three ways:

(1) *The plug-and-chug syndrome.* Many students, perhaps instinctively, seek to minimize exposure. Usually this is done by postponing any work on the assigned problems until late the night before the due date, then flipping rapidly through the textbook to find formulas in which to insert the data supplied in the statement of the problem. The authors and editors take great pains to make this process easy, but that is not always obvious to a hasty, drowsy student.

(2) *The just-the-right-data syndrome.* Almost never does a patient tell a physician exactly what the doctor needs to know for a diagnosis, nothing less and nothing more. Yet, by long-established custom, that is what is done in textbook problems. This deprives the student of the opportunity to practice two key aspects of any genuine problem-solving: *asking* "what do I need to know?" and *discerning* what is significant information.

(3) *The don't-know-how syndrome.* Studies in cognitive science show that even quite able students cannot solve problems only slightly different from those they have done before, *unless* they have a qualitative understanding. The usual textbook problems condition students to rely on a carefully structured context, to follow a safe path to the right answer. Guessing and qualitative reasoning is thereby discouraged. Students too often do not discover how much they can figure out on their own, the most gratifying and essential lesson.

Problem-solving should be playful fun! For years I have preached this to students, warned them about the nasty syndromes, given them some "real-life" problems as antidotes for the timidity induced by textbooks, and urged them to adopt the heuristic methods taught by Polya in his charming book, *How to Solve It*.

Here is my favorite little sermon: Someday, most of you will be regarded as *expert* in some domain. You will then find that people come to ask your advice not for what you *know*, but because they expect that as an expert you can *guess* better what nobody knows.

Sad to say, at present many students still succumb to one or more of these virulent syndromes. But now Dan Barouch has tackled these pestilences under the banner, *Stamp out the Syndromes!* This little book presents 150 problems, spanning the whole subject matter of general and inorganic chemistry, that do *not* involve numerical calculations. Each has a plausible "real-life" setting (at least by the criteria of soap operas). Some even have more than one correct answer. One of them (#31) has as yet no known correct answer, but the wrong ones are nonetheless instructive (as often so for frontier science). In all cases, "Hints" are provided, but relegated to the back of each chapter in order to encourage self-reliance. By avoiding the syndromes, this book aims to help students nurture latent talent for qualitative reasoning. This will surely strengthen also the ability to handle numerical problems, especially of the honest kind, so-called "word" problems that require understanding to set up calculations (rather than merely finding a predigested formula to crunch). Most importantly, the book fosters habits of thoughtful, self-generated questioning and analytical thinking. That is the chief goal of a college education.

Despite his tender age, Dan has already had the equivalent of several years of teaching chemistry at Harvard. He was superb as the Head Teaching Fellow for Chemistry 10, so I was delighted when Dan agreed to take on this long-contemplated project. I am delighted too with the result. Dan's evangelical zeal, verve, and dedication have been wonderful to behold. Earnest students of his book will learn valuable things about human as well as molecular chemistry.

We hope students and colleagues who use this book will send us comments, amendments, and suggestions, in anticipation of a new edition.

For grants in support of this project, we are grateful to the PEW Foundation and to the Associate Dean for Undergraduate Education at Harvard.

Enjoy!--and may the Force be with you!

Professor Dudley Herschbach

Chapter 1 — Introductory Concepts and Chemical Reactions

The Elements of Life
(*problem one*)

All living organisms are comprised mostly of carbon, nitrogen, oxygen, and hydrogen. Other important but rarer elements include sodium, magnesium, phosphorus, sulfur, chlorine, potassium, and calcium. In addition, other "trace" elements exist in a variety of different tissues and are also essential for life. To give one of many possible examples, the human thyroid gland (located in your neck) secretes an important molecule containing iodine. This molecule, known as thyroxine, is important for regulating many metabolic life processes. Activation of thyroxine to its most useful form requires the element selenium.

For each of the elements mentioned above, write out its chemical symbol. How many protons, neutrons, and electrons would an isolated atom of each contain?

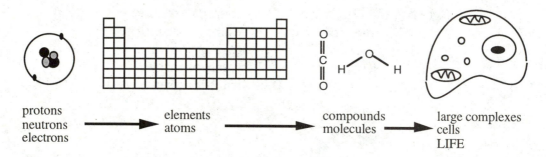

protons
neutrons
electrons

elements
atoms

compounds
molecules

large complexes
cells
LIFE

[category: introductory concepts and chemical reactions]
[topic: elements & compounds]
[difficulty: A]

Medical Chemicals
(*problem two*)

Two very simple compounds of nitrogen and oxygen, nitrous oxide and nitric oxide, have everyday uses in clinical medicine. Nitrous oxide, which is inhaled by a patient via a face mask, is used as a fast-acting general anesthetic for surgery. In contrast, nitric oxide is used to dilate coronary blood vessels in patients with angina (chest pain) or a suspected myocardial infarction (heart attack).

Due to its instability, nitric oxide cannot be used directly; instead, nitroglycerine tablets are admistered beneath the tongue. The nitroglycerine then enters the bloodstream and undergoes a reaction to produce nitric oxide in the patient's body. Nitric oxide is also produced normally by the body and has a wide variety of biological functions.

What are the chemical formulas of nitrous oxide and nitric oxide? What are the phases (solid, liquid, or gas) of the inhaled nitrous oxide and nitroglycerin tablets? Notice how similar these chemicals are, but with such different properties and uses!

[category: introductory concepts and chemical reactions]
[topic: elements & compounds]
[difficulty: A]

Name That Ion!
(problem three)

The nomenclature for polyatomic ions might seem complex, but several rules are generally followed. For example, consider a polyatomic ion series involving oxoanions (negatively charged complex ions containing oxygen). How would you name: Cl^-, ClO^-, ClO_2^-, ClO_3^-, ClO_4^-? What are the names of their corresponding acids: HCl, $HClO$, $HClO_2$, $HClO_3$, $HClO_4$?

How would you name the ions S^{2-}, SO_3^{2-}, SO_4^{2-}? How would you name H_2S, H_2SO_3, H_2SO_4?

[category: introductory concepts and chemical reactions]
[topic: elements & compounds]
[difficulty: B]

Metal Fires
(problem four)

What better place to start chemistry than with the first column of the periodic table? The alkali metals are very distinctive in their chemical properties. So distinctive that, in fact, they had been recognized as a group of chemically similar elements even before Mendeleyev's 1872 periodic table. They are all soft metals and form 1:1 compounds with the halogens and 2:1 compounds with oxygen. They also react with water to form hydrogen gas (which is explosive) and the corresponding alkali hydroxide. It turns out that the alkalis become more reactive as you go down this group. If you drop a piece of solid Na in water it reacts quickly and vigorously, whereas K reacts violently in a burst of flame as soon as the metal touches water, and Rb is so reactive it ignites spontaneously with water molecules in the air!

Write out the reaction of an alkali metal and water, making sure that it is balanced and that you include phases for each component.

[category: introductory concepts and chemical reactions]
[topic: chemical reactions]
[difficulty: A]

The Silver Tree
(problem five)

It's Christmas season, and you are looking for a tree. How much prettier, you think, it would be to have a tree with silver leaves instead of stringing tinsel around it. You set up a chemical reaction to make a "silver tree" by placing solid copper wire, in the shape of a tree trunk, in a solution of silver nitrate. You come back in an hour to find that the solution turned blue and that the copper wire has blossomed into a beautiful tree with countless numbers of silver leaves. Write the balanced chemical reaction for this process, and state why the solution turned blue.

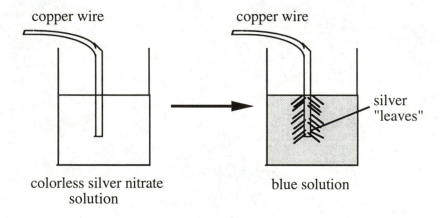

[category: introductory concepts and chemical reactions]
[topic: chemical reactions]
[difficulty: B]

Peroxide Potential
(problem six)

Hydrogen peroxide, H_2O_2, is a powerful multipurpose oxidant used in disinfection, bleaching, propulsion of rockets, and cleansing. You know that it can react completely with sulfuric acid and potassium permanganate to produce potassium sulfate, manganese (II) sulfate, water, and oxygen gas. Write down this process as a balanced chemical reaction.

[category: introductory concepts and chemical reactions]
[topic: chemical reactions]
[difficulty: B]

Don't Break the Law
(problem seven)

Your ten year-old niece wants to learn chemistry. You decide to teach her about the law of conservation of mass. "Matter cannot be created or destroyed," you begin, "no matter what you do."

To show her this, you take a gram of pure copper powder, weigh it, arrange it in a variety of shapes, and show her that its mass does not change. You then place it over a flame for a few minutes and, to your surprise, you find that its mass increased by about twenty-five percent!!

To save face (with your niece now laughing at you), you must explain what happened with the heated copper experiment. What is the fundamental difference between arranging the copper into a variety of shapes and heating? Be sure your explanation includes a chemical reaction. You can assume this reaction went to completion.

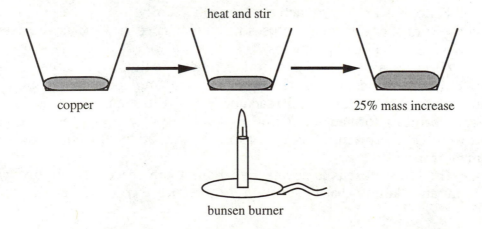

[category: introductory concepts and chemical reactions]
[topic: chemical reactions]
[difficulty: C]

Kaboom!
(problem eight)

Liquid nitroglycerin, a very powerful explosive, is a simple molecule with the formula $C_3H_5(NO_3)_3$. It is unstable and reacts to form carbon dioxide, nitrogen, water vapor, and oxygen. Write a balanced chemical reaction for the explosion of nitroglycerin, and state how many moles of gases are produced for each mole of nitroglycerin reacting. Expanding gases perform large quantities of work on the surroundings (ie, they can do a lot of damage). How does the volume occupied by a mole of a typical gas compare with a mole of a typical liquid? Then state why nitroglycerin is so explosive!!

[The expansion of the gases is only one of several factors that makes nitroglycerin very explosive. For example, the reaction proceeds very quickly because oxygen is contained *within* the liquid and thus does not have to "diffuse in". In addition, the reaction releases a large amount of heat.]

[category: introductory concepts and chemical reactions]
[topic: stoichiometry]
[difficulty: A]

Hectic Halogens
(problem nine)

Halogens are very reactive elements, readily form a wide variety of compounds, and have great uses in both science and industry. Here are two hectic halogen problems:

(i) Naturally occuring bromide ions in solution can react with molecular chlorine to form chloride ions and molecular bromine. If you start with a mole of each reactant, what is the maximum amount of each product formed? Assume that while performing the reaction you remove the bromine as it is formed, so that the reaction goes to completion.

(ii) Fluorine gas is so reactive that it can react with hydroxide ions in an alkaline (basic) solution to produce oxygen difluoride gas, fluoride ions, and water. Write out this chemical reaction, including stoichiometric coefficients and the phases of each reactant or product.

[category: introductory concepts and chemical reactions]
[topic: stoichiometry]
[difficulty: B]

It's on Fire!
(problem ten)

Your project is to determine the molecular formula of an unknown hydrocarbon (a compound consisting only of hydrogen and carbon). It is burned completely under tightly controlled laboratory conditions (i.e., all products are recoverable, there are no side reactions, etc.), and the only available source of oxygen is a balloon inflated with pure oxygen gas and attached through an inlet valve to the combustion chamber. A friend pulls you aside and whispers in your ear that the number of molecules in your hydrocarbon sample is the same as the number of oxygen molecules in the balloon. Keeping this tip in mind, you perform the combustion reaction and watch the balloon shrink. When the balloon is completely deflated, the combustion reaction stops, and you notice that four-fifths of the hydrocarbon remain.

What are the two possibilities for the molecular formula of the unknown hydrocarbon? Assume that you performed the experiment in a manner such that the combustion reaction essentially went to "completion", that all the hydrocarbon that reacted became carbon dioxide and water, and that there was little production of "soot" or other by-products.

Note: This problem is tricky but fun! Think about it for a while...

[category: introductory concepts and chemical reactions]
[topic: stoichiometry]
[difficulty: C]

Introductory Concepts and Chemical Reactions — HINTS

The Elements of Life
(problem one)

Hint 1: What are the relative masses of protons, neutrons, and electrons?

Hint 2: How is the atomic number of an element related to the numbers of the subatomic particles contained within each atom?

Hint 3: How is the atomic mass of an element related to the numbers of the subatomic particles contained within each atom?

Medical Chemicals
(problem two)

Hint 1: What are the possible oxidation numbers for nitrogen?

Hint 2: How are the "-ous" and "-ic" endings in a chemical compound related to oxidation states? Number of oxygen atoms?

Hint 3: What are the different phases of matter?

Name That Ion!
(problem three)

Hint 1: What is the "suffix" of the complex ion containing the most oxygens? Is it the same with oxoanions of Cl and S?

Hint 2: How are the "suffixes" of the complex ions altered to "suffixes" of their corresponding acids?

Hint 3: Is HCl an acid? H_2S?

Metal Fires
(problem four)

Hint 1: What is the chemical form of elemental alkali metals (monotomic, diatomic, etc.)?

Hint 2: What is the charge on the elemental alkali metals? The charge on the ionic alkali metals?

Hint 3: What is the chemical form of hydrogen gas (monotomic, diatomic, etc.)?

The Silver Tree
(problem five)

Hint 1: What happens to silver nitrate in solution?

Hint 2: Which aqueous ion has a characteristic blue color?

Hint 3: For Cu and Ag, what are the charges of the aqueous metal ions and the solid elemental metals?

Peroxide Potential
(problem six)

Hint 1: What are the formulas of the ions sulfate and permanganate?

Hint 2: It is often easier to balance the heavy elements first, and deal with the hydrogen and oxygen afterwards. Why is this true, and can it be used here?

Hint 3: What must be true for a balanced chemical reaction?

Don't Break the Law
(problem seven)

Hint 1: What is the fundamental difference between arranging the powder in a variety of shapes and burning it?

Hint 2: Explain what is happening when the copper is burned. Is it a physical or chemical change? If physical, explain where the extra mass came from. If chemical, write out the balanced reaction.

Hint 3: Can you identify the exact substance at the end of the burning based on the information in the question and thinking about the possible compounds of copper?

Kaboom!
(problem eight)

Hint 1: What are the phases and stoichiometric coefficients of each molecule in the reaction?

Hint 2: What is the volume of a mole of a typical gas (for example, an ideal gas at STP)? What is the volume of a mole of a typical liquid (for example, a compound with the same molar mass as nitroglycerin and the density of water)? Which is greater? By how much?

Hint 3: What is the relationship between volume expansion of a gas and explosive properties?

Hectic Halogens
(problem nine)

Hint 1: What are the natural molecular forms of the halogens?

Hint 2: What are the common ionic forms of the halogens?

Hint 3: How does the principle of mass balance influence the way you should write a chemical reaction?

It's on Fire!
(problem ten)

Hint 1: What is the general reaction for the combustion of a hydrocarbon (assuming that the combustion is complete)?

Hint 2: What's the limiting reagent? What's the stoichiometry of the reagents (hydrocarbon and oxygen) in the reaction?

Hint 3: From this information, what must be the stoichiometry of the products produced?

Chapter 2 — Gases

Problem	Title	Topic	Difficulty
11	Blowing Balloons	gas laws	A
12	Diving Deep	gas laws	A
13	Not Quite Perfect	gas laws	A
14	Lungs of a Smoker	gas laws	B
15	Filled with Hot Air	gas laws	B
16	A Candlelight Moment	gas laws	B
17	The Grinch	gas laws	B
18	The Fountain of Blood	gas laws	C
19	Frigid Nectar	gas laws	C
20	Under Pressure	gas laws	C
21	Above the Clouds	kinetic theory of gases	A
22	Molecules in Motion	kinetic theory of gases	A
23	A Sense of Pitch	kinetic theory of gases	B
24	Growing Bubbles	kinetic theory of gases	B
25	Gaseous Philosophy	kinetic theory of gases	C

Hints   **25**

Blowing Balloons
(problem eleven)

You have three identical balloons at room temperature. You inflate each one to the same size, filling them with different gases. You put helium in balloon I, nitrogen in balloon II, and xenon in balloon III (see illustration). You realize that the pressure inside each balloon is equal to atmospheric pressure. Why? You notice that balloon I floats to the ceiling, balloon II feels light, and balloon III feels heavy. Which balloon contains the most gas molecules? What accounts for the differences in weight? Assume all gases are ideal.

You then take balloon II and throw it into a bucket of liquid nitrogen. It deflates. Why did this happen? If you were to weigh the inflated balloon and the deflated balloon on a scale, which would weight more? Which would contain more nitrogen molecules?

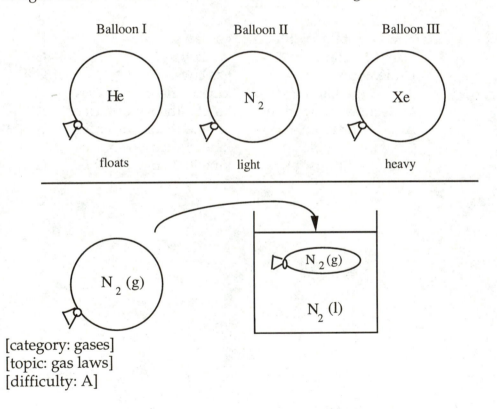

[category: gases]
[topic: gas laws]
[difficulty: A]

Diving Deep
(problem twelve)

You are a scuba diver deep below sea level, and you are acutely aware of the dangers of ascending back to the surface. First, you know that your ascent must be slow in order to avoid "the bends" (a property of gas solubility in liquids that is explored in one of the "solids and liquids" problems in chapter 3). Second, you know that you must exhale a larger volume than you inhale during the ascent, and if you don't (say if you hold your breath) your lungs might rupture. Explain this second effect using your knowledge of the ideal gas law.

[category: gases]
[topic: gas laws]
[difficulty: A]

<u>Not Quite Perfect</u>
(problem thirteen)

You have finally learned the ideal gas law. You pick your favorite gas (say, methane) and decide to test it. You apply high pressures at room temperature and measure the corresponding changes in volume. You then do some calculations and plot PV/nRT on the y-axis versus pressure P on the x-axis. You expect the graph to be just a straight horizontal line, y=1 (curve A). To your surprise, you get curve B instead. You scratch your head and try to think of why your gas is nonideal. What is the major factor contributing to nonideality at the lower pressures (point I)? At the higher pressures (point II)?

You then repeat your experiment but measure the pressure and volume at a very high temperature. This yields a somehat more ideal curve (curve C). Why does high temperature favor ideality?

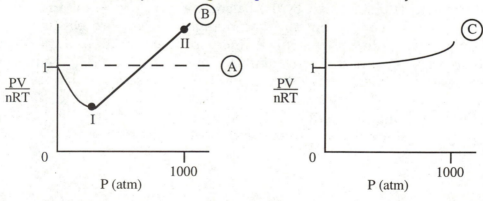

[category: gases]
[topic: gas laws]
[difficulty: A]

Lungs of a Smoker
(problem fourteen)

This problem will demonstrate the application of the ideal gas law in a common medical situation. You are a physician and have a patient whom you suspect has an abnormal lung "functional residual capacity" (FRC). The FRC is defined as the volume of air in the lungs after a passive exhalation. A patient's FRC is increased in certain types of pulmonary diseases. One such disease is emphysema, which often results from long term tobacco smoking. You want to determine your patient's FRC in order to determine the degree of impairment of your patient's lung function.

You have a tank (volume V_1) containing 1 atm of a mixture of gases (essentially air) including helium at an initial concentration C_1. This tank has one small opening, and a short tube connects the tank with a mouthpiece. You place the mouthpiece in your patient's mouth at the end of a passive exhalation, and then the patient begins breathing into the tank. The patient performs several inhalations and exhalations into the tank. You measure the concentration of He in the tank at various points in time. You notice that the concentration drops fairly quickly but stabilizes at a lower concentration C_2 (see illustration on next page). Explain the observation, and describe how this can lead to an estimate of the patient's FRC. Note that He is an inert gas and is relatively insoluble in blood. (Assume that none is taken up into the bloodstream from the lungs.)

[Note for those interested in physiology or medicine: The natural FRC is determined by the balance of two forces— the tendency of the chest wall to "spring out" and the natural "inward elastic recoil" of the lungs. With heavy smokers, there is a deterioration of the elastic membranes in the lung leading to emphysema. The lungs of emphysemic patients have less inward elastic recoil, and thus their FRC increases. The degradation of the elastic lung membranes impairs pulmonary function, particularly the exchange of oxygen and carbon dioxide that normally occurs across these alveolar membranes.]

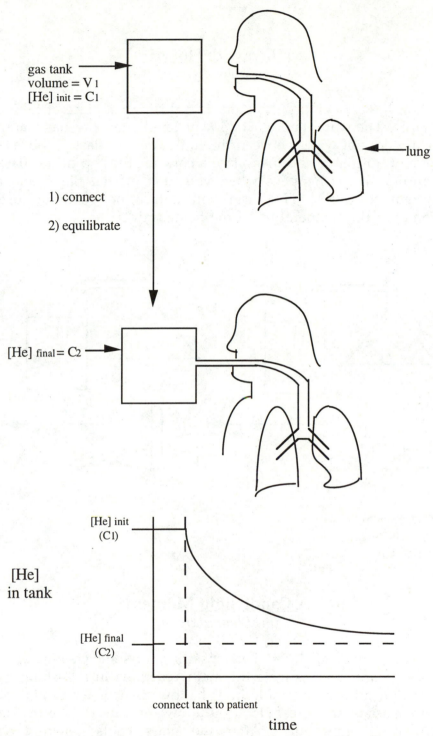

gas tank
volume = V_1
$[He]_{init} = C_1$

lung

1) connect

2) equilibrate

$[He]_{final} = C_2$

[He] init
(C_1)

[He]
in tank

[He] final
(C_2)

connect tank to patient

time

[category: gases]
[topic: gas laws]
[difficulty: B]

Filled with Hot Air
(problem fifteen)

You have an empty flask attached through a piece of tubing to a water tank. The water tank is at a lower level than the flask, and thus there is no flow of water between the tank and the flask. You heat the flask over a hot flame and then place a cork on the top of the flask (see illustration). What happens when you turn off the flame and allow the system to cool? How can you determine, without using a thermometer, the temperature of the heated air?

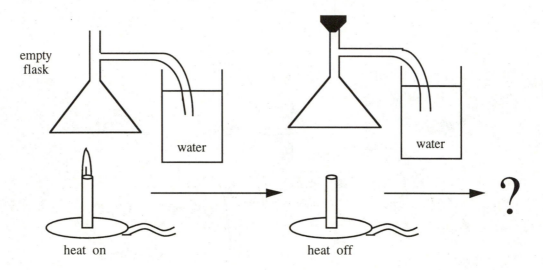

[category: gases]
[topic: gas laws]
[difficulty: B]

A Candlelight Moment
(problem sixteen)

Dining by candlelight. Perfect setting for some chemistry.
You want to investigate the main reactions in a burning candle. You position a burning candle in a shallow dish of water and place a tall, thin glass over the candle. You allow the candle to burn until it spontaneously goes out. After the glass cools down to room temperature, you notice that the level of water in the glass has risen slightly. You then take the glass, quickly place it upright, add some

limewater (calcium hydroxide solution), and cover it. After swirling the liquid in the glass, you notice the appearance of a white precipitate.

Explain all the observations during the candle burning. You can assume that the combustion reaction went to completion. Why did the candle go out? Why did the water height rise? Is the pressure inside the glass constant? If so, what is the pressure? What is the composition of the gas before adding the limewater? What is the white precipitate?

[category: gases]
[topic: gas laws]
[difficulty: B]

The Grinch
(problem seventeen)

You are a hoarder and miser (and thus must suffer eternally in Dante's Inferno). You absolutely cannot *stand* buying anything when not absolutely necessary. For some reason, you especially hate buying balloons. You have to provide the balloons for both your Christmas and New Year's parties. You cleverly buy a bag of ten balloons with the writing "Merry Christmas and Happy New Year!" Great! You can use the same balloons for both occasions. You blow up the balloons, throw a wonderful Christmas party, and then are grief-stricken to read the fine print on the balloon bag:

"Balloons will remain inflated approximately four days. Too bad for all you misers."

What can you do to extend the lifetime of the balloons to the New Year? You may use any common household item/appliance. Unfortunately, you cannot by any means undo the knot at the base of the balloon or re-inflate the balloon.

[category: gases]
[topic: gas laws]
[difficulty: B]

The Fountain of Blood
(problem eighteen)

Waterfalls and fountains have long been thought to possess religious and supernatural properties. They have been the goal of explorer's quests, the centerpiece for cities, and the power of nature in visual form. The mystique and magic of such radiant displays of nature's beauty are admired by everyone.

You fill a large flask with water and a few milliliters of the indicator phenolphthalein. On a ring stand, you position a large, inverted, round bottom flask filled with pure ammonia gas directly above the first flask. This flask is stoppered in order to prevent the loss of any ammonia, and the ammonia is colorless. You then connect the two flasks by a long glass tube, filled with water along its length, that reaches from the water reservoir, through the stopper, and into the top portion of the inverted flask filled with ammonia (see illustration).

You watch closely. Nothing happens for a few seconds. Then the liquid at the top of the glass tube becomes red and gently overflows into the top flask. More and more liquid begins to spout out of the glass tube and turn red. The drama continues and climaxes as a huge amount of colorless water turns crimson as it fountains into the top flask. You've made a fountain of blood!!

Explain why you get an accelerating "fountaining" effect in this experiment, why the water flows up against gravity, and why it turns red as it spouts out of the glass tube.

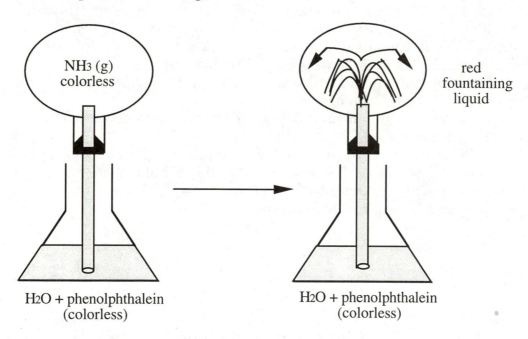

NH₃ (g) colorless

red fountaining liquid

H₂O + phenolphthalein (colorless)

H₂O + phenolphthalein (colorless)

[category: gases]
[topic: gas laws]
[difficulty: C]

Frigid Nectar
(problem nineteen)

It's a brisk December day in Cambridge, and you find a styrofoam container of some strange liquid, packed in a well-insulated box, on your front porch. It looks really cold, and you wonder what its temperature is. Sadly, you find that your thermometer has been broken, and you run to your neighbor's house, doing your best to dodge the puddles that have lined the sidewalk since last night's rainstorm. You see that the puddles are just barely beginning to freeze. Winter's finally here, you think. Shivering, you quickly reach your neighbor and ask for his thermometer. He does not have one and can only offer you a piece of string, a straight ruler, and a small balloon. Using nothing else, describe, both qualitatively and quantitatively, how you could *very accurately* determine the temperature of your strange liquid.

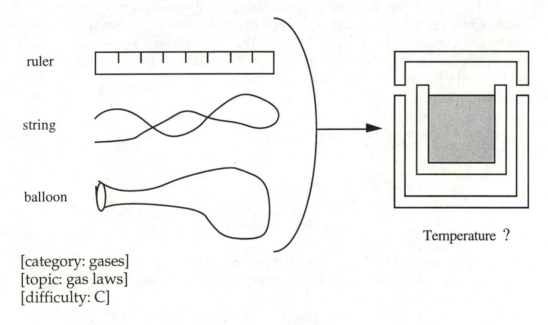

Temperature ?

[category: gases]
[topic: gas laws]
[difficulty: C]

Under Pressure
(problem twenty)

You have a sample of a pure gas in a rigid-walled vessel. This compound consists only of nitrogen and oxygen atoms. The complete decomposition of this gas converts it to molecular nitrogen and molecular oxygen. After the gas has decomposed completely at a constant temperature, you notice that the pressure has risen by about fifty percent. What could be the identity of your original gas?

[category: gases]
[topic: gas laws]
[difficulty: C]

Above the Clouds
(problem twenty-one)

You are soaring above the clouds in an airplane on the way to a tropical island. You are grateful for the comfortable pressurized cabin air, since you know that outside the cabin the pressure of the air is much less than atmospheric pressure at sea level. You decide to compare the outside air "above the clouds" and "on the ground".

For all of the following quantities, state whether they would be greater for the air above the clouds or on the ground: density, distance between molecules, mean free path, molecular speed, collision frequency, kinetic energy per molecule. Assume that the temperature is the same above the clouds and on the ground.

[category: gases]
[topic: kinetic theory of gases]
[difficulty: A]

Molecules in Motion
(problem twenty-two)

Here are a few shorties on gas laws and the kinetic theory of gases! They shouldn't take more than a few seconds each. Ready at your buzzers? Go!

(i) Circle the greater quantity or state if they are equal: the average molecular speed of gas molecules or the rate of effusion of a gas

(ii) If a gas at room temperature and atmospheric pressure has a density d, then what happens to d if you double the pressure (in atmospheres)? If you double the temperature (in Celsius)?

(iii) Which has a higher molecular velocity at room temperature and atmospheric pressure, helium or argon? Which has a higher kinetic energy?

[category: gases]
[topic: kinetic theory of gases]
[difficulty: A]

A Sense of Pitch
(problem twenty-three)

Perhaps you would like to be a soprano and a bass in the same evening. With the aid of a few gases and a knowledge of chemistry, you can achieve this feat. Why would inhaling helium make your voice become much higher than normal? Why would inhaling sulfur hexafluoride make your voice much lower than normal? Explain in terms of the kinetic theory of gases. Would your voice be higher or lower than average if you're in the hospital and placed in an oxygen-rich environment?

[category: gases]
[topic: kinetic theory of gases]
[difficulty: B]

Growing Bubbles
(problem twenty-four)

You can create bubbles by placing a wand with a loop in a solution of soapy water and waving the wand in the air. You notice that these bubbles are rather small and can float on a beaker that is filled with carbon dioxide. Upon close examination, however, you notice that the bubbles floating on the carbon dioxide seem to grow!

According to the kinetic theory of gases, which would effuse faster, air or carbon dioxide? Why is the observed result surprising? You must conclude that the bubble growth is not solely due to simple effusion of gas molecules. There is a "semipermeable" barrier that separates the interior of the bubble from its environment. Based on the observation, how do you suspect this membrane allows the relative transport of nitrogen, oxygen, and carbon dioxide? Would bubbles grow more quickly or more slowly if you used *more* soap in your initial soapy water mixture?

[category: gases]
[topic: kinetic theory of gases]
[difficulty: B]

Gaseous Philosophy
(problem twenty-five)

You are comparing the behavior of xenon with sulfur hexafluoride. You have one sample of each gas, but somehow these lack labels. Use the following information in order to devise an experiment that could accurately determine which gas is which.

Roughly speaking, xenon and sulfur hexafluoride are about the same molecular mass, and thus you cannot make accurate measurements about their relative effusion rates. However, certain differences make them have very different properties. Which one would be expected to be more ideal in nature? Which would have more internal energy at room temperature? Which has a higher heat capacity (Cv)? Why?

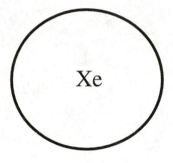 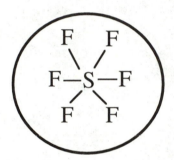

[category: gases]
[topic: kinetic theory of gases]
[difficulty: C]

Gases — HINTS

Blowing Balloons
(problem eleven)

Hint 1: What is the ideal gas law?

Hint 2: What is the relationship between moles and mass of a gas?

Hint 3: What is the relationship between moles and number of molecules of a gas?

Diving Deep
(problem twelve)

Hint 1: What is the ideal gas law? Specifically what is its pressure-volume relation?

Hint 2: What sorts of pressures exist under water? At the surface?

Hint 3: Your lungs are like balloons- they can be filled up to varying amounts. What happens if you keep blowing up a balloon?

Not Quite Perfect
(problem thirteen)

Hint 1: What is van der Waals equation of state?

Hint 2: Are gas molecules really points in space that have no volume and do not interact with each other?

Hint 3: At a higher temperature, which nonideality factor predominates?

Lungs of a Smoker
(problem fourteen)

Hint 1: What is the ideal gas law? What is the relationship between volumes and concentrations of gases?

Hint 2: How can the setup be considered a "two-vessel" system?

Hint 3: Why is it important that the He be inert and insoluble?

Filled with Hot Air
(problem fifteen)

Hint 1: What happens to the molecules of air when the gas is heated?

Hint 2: When the flask is corked and cooled, which variables in the system are fixed? What variable is allowed to change, and what must be the response of the system?

Hint 3: What's the relationship between temperature and volume of a gas? How can this be used to calculate the temperature of the hot air?

A Candlelight Moment
(problem sixteen)

Hint 1: What is the chemical composition of candlewax? What is the combustion reaction that produces the flame?

Hint 2: Can you use the gas laws to explain the observations?

Hint 3: Limewater would not form a white precipitate when mixed with air. What was responsible for forming the precipitate?

The Grinch
(problem seventeen)

Hint 1: What is the physical cause of a balloon deflating?

Hint 2: According to the kinetic theory of gases and gas laws, what factors determine how fast a balloon will deflate?

Hint 3: What common household items/appliances can control these variables?

The Fountain of Blood
(problem eighteen)

Hint 1: What is chemical reaction of gaseous ammonia and liquid water?

Hint 2: What is phenolphthalein and how is it used as a pH indicator? Why is the liquid in the bottom flask colorless but the liquid in the top flask red?

Hint 3: Why does more water rising in the glass tube accelerate this fountaining process?

Frigid Nectar
(problem nineteen)

Hint 1: How does the balloon volume relate to its temperature?

Hint 2: How can a ruler and a piece of string be used to measure the volume of a sphere?

Hint 3: What's the temperature outside?

Under Pressure
(problem twenty)

Hint 1: Which variables are constant in this system?

Hint 2: What are the products of the decomposition? What is the relation between products and reactants?

Hint 3: What is the relation between pressure and stoichiometry?

Above the Clouds
(problem twenty-one)

Hint 1: What is the ideal gas law? How can it be rearranged to yield an equation containing the relationship between density and pressure?

Hint 2: In a given volume of gas at constant temperature, would there be more particles at sea level or at a high altitude?

Hint 3: What variables determine molecular speed and kinetic energy for an ideal gas?

Molecules in Motion
(problem twenty-two)

Hint 1: How is the relationship between molecular speeds (microscopic quantity) and effusion rates (macroscopic quantity) analogous to the relation between displacement and total distance travelled in a random walk?

Hint 2: How can the ideal gas equation be rearranged to give a formula for density of a gas? Which variables does the density depend on?

Hint 3: At constant temperature and pressure, what variables affect kinetic energy of an ideal gas? Molecular speed?

A Sense of Pitch
(problem twenty-three)

Hint 1: What are the molar masses of helium, air, and sulfur hexafluoride?

Hint 2: Given a constant external temperature, which gas would travel the fastest and slowest?

Hint 3: Which average molar mass is higher, normal air or oxygen-enriched air?

Growing Bubbles
(problem twenty-four)

Hint 1: What is the average molar mass of air? Carbon dioxide? Which is heavier?

Hint 2: What is Graham's law of effusion? What chemical reactions or physical processes must occur for oxygen, nitrogen, or carbon dioxide to pass through the soapy water membrane?

Hint 3: To what extent do these reactions occur? Do they occur more when the soapy water has a lower or higher pH?

Gaseous Philosophy
(problem twenty-five)

Hint 1: What is the main difference between xenon and sulfur hexafluoride? How many degrees of freedom does each gas have?

Hint 2: What is the definition of an ideal gas? What assumptions are made for ideality?

Hint 3: What is the definition of heat capacity (Cv)? How does it relate to the energy of a gas at a given temperature?

Chapter 3 — Solids and Liquids

Get It Together!
(problem twenty-six)

The theory of ideal gases postulates that gas molecules are points in space that do not attract or repel each other. In solids and liquids, however, such intermolecular forces are vital to holding a substance together. These forces give rise to the characteristic properties of solids and liquids, such as possessing a definite volume, having a shape (solids), exhibiting surface tension (liquids), etc. What are the relative strengths of the intermolecular forces in solids, liquids, and gases? How do intermolecular forces qualitatively relate to melting and boiling points? Which quantity is greater, the heat of fusion or the heat of vaporization of a substance? Why?

There is a wide range of melting points among solids. For example, consider the solid forms of the following: sodium chloride, diamond, ice (solid water), carbon dioxide, calcium oxide, helium, and hydrogen sulfide. What types of forces hold each of these solids together? Order these compounds in terms of increasing melting point.

[category: solids and liquids]
[topic: intermolecular forces]
[difficulty: A]

Name That Molecule
(problem twenty-seven)

You have three refrigerated vessels each containing different organic liquids, all of which are hydrocarbons of the formula C_5H_{12} (see illustration on next page). Since these compounds have the same chemical formula but are structurally different, they are called "isomers".

Sadly, your three vessels are all unlabeled. Unfortunately, you have no handy chemistry reference book to look up their physical properties. You only have a thermometer, glassware, and a heat source. Explain how you can determine which is which. (Think about the differences in the intermolecular forces involved.)

```
      H  H  H  H  H                      H                         H
      |  |  |  |  |                  H— C—H                    H— C—H
  H— C—C— C— C— C—H          H        |        H          H  H  |  H
      |  |  |  |  |          |        |        |          |  |  |  |
      H  H  H  H  H      H— C —— C —— C —H   H— C—C— C— C—H
                             |        |        |          |  |  |  |
                             H        |        H          H  H  H  H
                                  H— C—H
                                      |
                                      H
```

    C_5H_{12} (Isomer #1) C_5H_{12} (Isomer #2) C_5H_{12} (Isomer #3)

[category: solids and liquids]
[topic: intermolecular forces]
[difficulty: B]

Hairy Hydrogen
(problem twenty-eight)

Hydrogen bonding is an extremely important force in biochemistry and for any reactions that take place in an aqueous environment. It is an intermolecular force involving a hydrogen atom bridged between two extremely electronegative atoms, usually oxygen, nitrogen, fluorine, or chlorine. The hydrogen is covalently bonded to one atom and hydrogen bonded to the other. The more electronegative the atoms flanking the H, the stronger is the hydrogen bond. It is much stronger than standard dipole-dipole interactions but weaker than a covalent bond.

(i) Explain why water has a much higher boiling point than hydrogen disulfide, even though both are similar in structure and the latter has much greater London forces. Which forces predominate in each compound?

(ii) Describe why water has a much higher boiling point than hydrofluoric acid (HF). Note that both comopunds are polar and will form hydrogen bonds, and that an F-H-F hydrogen bond is stronger than an O-H-O hydrogen bond because fluroine is more electronegative than oxygen.

(iii) Now that you're warmed up, here is a more challenging hairy hydrogen bonding problem. Assume you have a sample of a compound $C_6H_6O_2$, which is a benzene ring with two attached hydroxyl groups. This chemical formula can describe three different compounds with different properties, depending on the relative positions of the hydroxyl groups (see illustration). Think about the hydrogen bonds and intermolecular forces, and then order them in terms of increasing melting points. Explain your reasoning.

Hydrogen bonds:

$C_6H_6O_2$ Structures:

1.

2.

3.

[category: solids and liquids]
[topic: intermolecular forces]
[difficulty: B]

Purely Confusing
(problem twenty-nine)

You have in a beaker some very cold liquid that consists of only C_2H_6O molecules. You know that when a pure substance undergoes phase transitions, the temperature of the mixture should remain constant. (Thus a mixture of ice and liquid water will be zero degrees Celsius, and transferring heat into the mixture will first melt the ice before raising its temperature.) You decide to experiment with your C_2H_6O by heating it. You notice that about half is vaporized quickly while the other half remains liquid. The temperature then rises steadily until a much higher temperature, at which point the rest vaporizes (see illustration). What could your mysterious liquid be? Explain the unexpected boiling results.

Heating H_2O

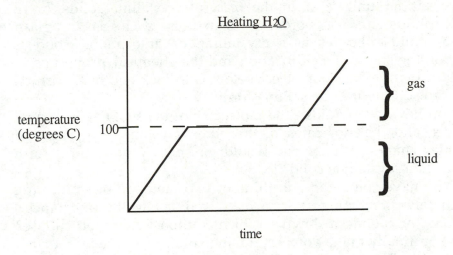

Heating C_2H_6O (Your beaker)

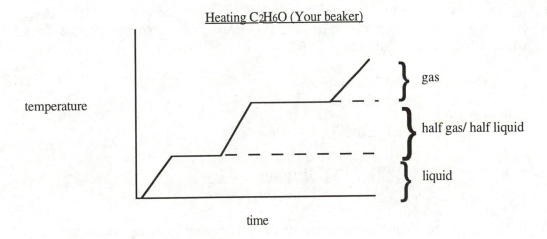

[category: solids and liquids]
[topic: intermolecular forces]
[difficulty: C]

Crystals and Glass
(problem thirty)

Consider the forces holding a solid together. In a solid, molecules, atoms, or ions are held together tightly, either in ordered arrays of repeating "unit cells" in the case of crystalline solids, or in more amorphous and less-ordered structures such as in glass. Whether a liquid solidifies as a crystalline or amorphous solid is dependent on the speed of solidification and the chemical properties of the substance involved. Your object below is to pick one of the given liquids and to solidify it in crystalline form!

(i) You have two liquids - the molecules of the first one are small and compact, and the molecules of the other substance are long and have complex shapes. Which one would you pick to have more of a chance to form a crystalline solid?

(ii) You have your solid and think of two methods of solidifying it - immediately immersing it in a bath well below the substance's freezing point, or cooling it slowly. Which method would you choose?

(iii) You finally make your crystalline solid. You are told that it is a cubic lattice, but you do not know if it is a simple cubic, body-centered cubic, or face-centered cubic lattice. For each of the three cases, how many molecules per unit cell and how many nearest neighboring molecules would there be?

[category: solids and liquids]
[topic: intermolecular forces]
[difficulty: C]

Gliding on the Ice
(problem thirty-one)

You are about to compete in the Olympics as a figure skater. You know that the blades on your skates are necessary for you to glide gracefully across the ice. You also know that you leave a path behind you where your blades cut into the ice. Five minutes before you enter the rink, you face a problem. You remember that your scientist friend had once told you that there is a lot of friction between metal and ice, yet you feel that there is little friction when you twirl and spin in the rink. Thus skating cannot simply be "cutting" the ice or melting the ice from the heat generated by the friction. You wonder how it is that you can perform this seemingly impossible feat. Come up with a plausible explanation for this dilemma soon, before you skate in front of the world! Think about the pressure, temperature, and phase changes involved. You may refer to the schematic phase diagram for water illustrated below.

[Note that the "plausible" explanation that you and many texts will use to explain this dilemma is actually technically incorrect. A issue of *Nature* several years ago shows that the "gliding" effect cannot be due to any simple explanations. The "actual" explanation of this phenomenon is unknown!!]

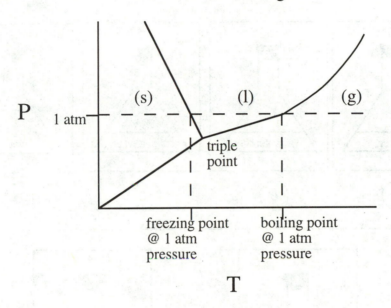

H2O Phase Diagram

[category: solids and liquids]
[topic: phase equilibria]
[difficulty: A]

Vapor Voodoo
(problem thirty-two)

Here are two shorties to review the concepts of vapor pressure and colligative properties:

You have four identical empty flasks. Into flasks A and B you add a small quantity of distilled water. Into flask C you add the same volume of seawater. Flask D you fill nearly completely with distilled water. You then place stoppers on the tops of all the flasks, and you leave all the flasks at room temperature except for flask B which you place at a cooler temperature (see illustration). After a short period of time you measure the vapor pressures of all four flasks. Which would you predict to have the higher vapor pressure, A or B? A or C? A or D? Explain.

You now take two evacuated flasks. You have a large travel budget, and so you fill one with air from Boston (very humid) and one with air from LA (very dry). Boston and LA were the same temperature for the days you collected the air, and both flasks are at one atmosphere pressure. Which flask has more nitrogen and oxygen molecules?

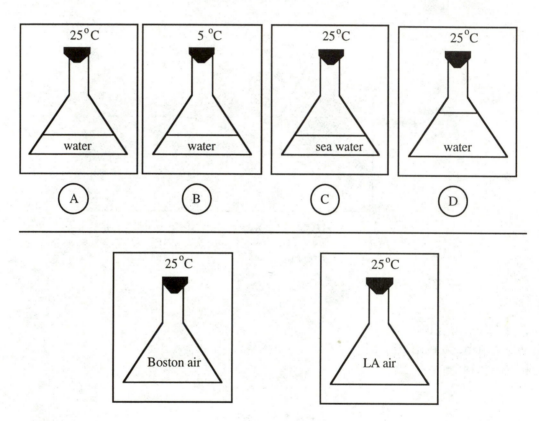

[category: solids and liquids]
[topic: phase equilibria]
[difficulty: A]

Cliffhanger's Cooking
(problem thirty-three)

You are an avid mountain climber and are heading for the most dangerous cliffs in the world. You also love to cook hard boiled eggs, and thus always carry a battery-operated hot pot and a dozen eggs on all your expeditions. When you reach the summit, you turn on the hot pot, wait for the water to boil, and then drop an egg in and time how long it takes to cook it to perfection. You're really fussy about getting the egg well done but not overdone, and thus you must be very precise about how long it takes to cook the egg. To your amazement, the time for cooking an egg to perfection is slightly different on different mountain peaks, and these measurements are very different from the cooking time in your house which is in a valley. Explain qualitatively why this is the case.

[category: solids and liquids]
[topic: phase equilibria]
[difficulty: B]

Enriching Thoughts
(problem thirty-four)

In your last trip to the mysterious shrines of Central Asia, you heard a prophet tell you an enriching thought — that the vapor pressure of benzene exceeds the vapor pressure of toluene at any given temperature. Now you are back home, and you have before you a sealed vessel containing an equimolar solution of benzene and toluene. You collect the vapor that is produced and recondense it in another vessel. What would be (qualitatively) the relative amounts of these two components in the liquid of the first vessel, in the vapor of the first vessel, in the liquid of the second vessel, and in the vapor of the second vessel? Draw a two-component phase diagram to illustrate these relative quantities, assuming that the mixture is an ideal solution.

[category: solids and liquids]
[topic: phase equilibria]
[difficulty: B]

Boiling by Cooling
(problem thirty-five)

You fill a flask about a tenth full with water and heat it over a flame. You boil the water rapidly for a few minutes and then quickly place a stopper over the top of the flask. You remove the flask from the flame, and after a few seconds the water in the flask stops boiling. You then place the sealed flask under cold tap water, and behold! The water starts boiling again!! Make the tap water even colder and the water in the flask boils faster!!!

Explain in detail what's going on. Don't believe that this will happen? Just try it.

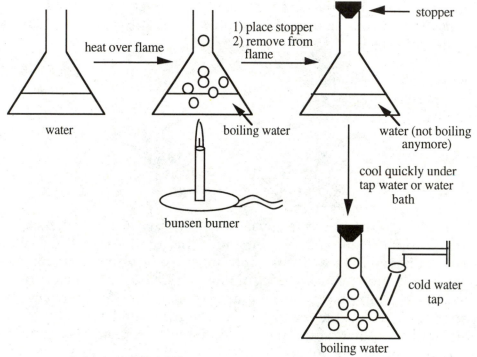

[category: solids and liquids]
[topic: phase equilibria]
[difficulty: C]

Jingle Bells, Jingle Bells
(problem thirty-six)

While singing the "Jingle Bells" tune, determine which salt you should use to melt the ice most effectively (on a mole per mole basis):

Dashing through the snow, in a one horse open sleigh
O'er the fields we go, crying all the way!
Slipping on the ice, I hate to watch kids play
So which salt shall I use to melt the ice today?
Oh —
Jingle bells, chem lab smells, and pick a salt with pride
Sodium, aluminum, or calcium chlori-ide!
Jingle bells, evil swells, the grinch is here to stay
So why don't you just give in now and ruin the ice today!

[category: solids and liquids]
[topic: colligative properties]
[difficulty: A]

New Year's Eve
(problem thirty-seven)

You're planning a New Year Eve's party, and for your guests you want nothing but the best. But then you realize that the Champagne is not chilled. Quickly, you put several bottles of unopened Champagne in your cooler that is just below zero degrees Celsius. You are so frenzied that you remember the Champagne only hours later. Should you be worried that the Champagne might be frozen? How about if you also left a bottle of pure, filtered water in the cooler? Explain.

[category: solids and liquids]
[topic: colligative properties]
[difficulty: A]

Scuba and Soda
(problem thirty-eight)

Henry's Law states that the solubility of a gas in a liquid is proportional to the partial pressure exerted on the liquid. Use this relationship to explain two familiar phenomena:
(i) When scuba divers are deep beneath the surface and decide to rise, they must rise very gradually. If they rise too quickly, then nitrogen gas bubbles out of their blood, causing extreme pain and danger. This is known as the "bends".

(ii) When a bottle of Coke is sealed, you do not see any bubbles forming and the soda does not become flat. When the bottle is opened, carbon dioxide bubbles out of solution, creating what we call "fizz". If you leave an open bottle of Coke on the table, then soon it will become flat.

[category: solids and liquids]
[topic: colligative properties]
[difficulty: B]

Crazy Colligatives!
(problem thirty-nine)

Determine the greater quantity in each case!

(i) Substance A has a larger molar mass than substance B. You dissolve a gram of each in water. Which solution has a higher vapor pressure?

(ii) In the solutions made above, which solution has a higher boiling point? Which has a higher freezing point?

(iii) You have two solutions, one containing one molar sugar and the other containing one molar table salt (sodium chloride). Which would have a higher osmotic pressure?

(iv) You have two red blood cells. You dump one in pure water and dump the other in serum (the liquid component of blood after the cells have been removed). In which case would there be a larger osmotic pressure difference?

[category: solids and liquids]
[topic: colligative properties]
[difficulty: B]

Cold Beer Anyone?
(problem forty)

You are having some friends over and so you go the store to buy a few cans of beer. Unfortunately the only beer available is at room temperature. Not a fan of warm beer, you put the cans in the freezer for about an hour to cool them. You take them out and shake them to make sure that they have not frozen. Luckily, they are very cold and still liquid. You then open one can to drink— but as soon as you open it the beer freezes!!

Propose a reasonable explanation for why the liquid beer froze when the can was opened.

[category: solids and liquids]
[topic: colligative properties]
[difficulty: C]

Solids and Liquids — HINTS

Get It Together!
(problem twenty-six)

Hint 1: What are the general structures of crystalline solids, liquids, and ideal gases? Which are more similar in terms of intermolecular forces, solids and liquids, or liquids and gases?

Hint 2: Which forces are the strongest? Consider ionic, covalent, dipole, and London forces.

Hint 3: When are ionic forces the strongest? When are dipoles the strongest? When are London forces the strongest?

Name That Molecule
(problem twenty-seven)

Hint 1: What are the molecular masses of each isomer?

Hint 2: How do the three different C_5H_{12} compounds differ? How does this affect their intermolecular forces?

Hint 3: What physical property can you measure with a thermometer and heat source?

Hairy Hydrogen
(problem twenty-eight)

Hint 1: What makes hydrogen bonds so strong?

Hint 2: Observe the Lewis structures of water and HF. How do the molecular geometries explain why water has the higher boiling point?

Hint 3: How do the relative positioning of the hydroxyl groups on the benzene ring account for the different melting points?

Purely Confusing
(problem twenty-nine)

Hint 1: Why are these boiling results unexpected?

Hint 2: What do these results say about the purity of your initial sample? What explanation can resolve the facts that your liquid boils over a large temperature range yet consists of only C_2H_6O molecules?

Hint 3: What sort(s) of intermolecular force(s) must be present in your unknown liquid? How can they be used to help explain the boiling results?

Crystals and Glass
(problem thirty)

Hint 1: Which would "pack" better in a lattice, small compact molecules or large complex ones?

Hint 2: The molecules in a solid can adopt many conformations, few of which are good crystals. Does this imply that an abrupt or gradual time of solidification would be better?

Hint 3: What are the cardinal features of sc, bcc, and fcc lattices?

Gliding on the Ice
(problem thirty-one)

Hint 1: Study the phase diagram of H_2O (see illustration). Where are the regions and dividing lines between solid, liquid, and gas?

Hint 2: Because there is a lot of friction between metal and ice, and close contact between the skater and the ground, what must the interface be?

Hint 3: How can this state be achieved? Describe using phase diagram shifts as well as the physical reality of a skater on the ice.

Vapor Voodoo
(problem thirty-two)

Hint 1: What determines vapor pressure of a pure substance?

Hint 2: How can Raoult's Law explain how impurities alter vapor pressure?

Hint 3: What is the difference between humid and dry air?

Cliffhanger's Cooking
(problem thirty-three)

Hint 1: What causes boiling (in terms of phase equilibria)? Is 100°C always the condition for boiling water?

Hint 2: What variable is different between mountain peaks and your house?

Hint 3: How does physical location affect cooking time?

Enriching Thoughts
(problem thirty-four)

Hint 1: Which component is more volatile, toluene or benzene?

Hint 2: How would Raoult's Law help determine what is in the vapor above a liquid solution?

Hint 3: How could this process be generalized into a purification/enrichment procedure (fractional distillation)?

Boiling by Cooling
(problem thirty-five)

Hint 1: What are the contents of the flask before boiling? After boiling?

Hint 2: Why is sealing the flask essential to the experiment?

Hint 3: What does the cooling of the flask have to do with the water re-boiling? Under what conditions does a liquid boil?

Jingle Bells, Jingle Bells
(problem thirty-six)

Hint 1: What are the empirical formulas of sodium chloride, aluminum chloride, and calcium chloride?

Hint 2: What factors determine how effectively salts melt ice? Think about colligative properties.

Hint 3: Can you estimate quantitatively the relative effectiveness of the three salts on a per mole basis?

New Year's Eve
(problem thirty-seven)

Hint 1: What is the freezing point of water?

Hint 2: What is the approximate composition of Champagne?

Hint 3: How do solution composition and pressure influence freezing point?

Scuba and Soda
(problem thirty-eight)

Hint 1: What is the mathematical relationship between solubility and pressure?

Hint 2: What is the composition of a person's blood at sea level and deep beneath the surface of the water?

Hint 3: What variable is being changed when you open the bottle of Coke?

Crazy Colligatives!
(problem thirty-nine)

Hint 1: A gram of which substance would yield a higher molality, one with a high or low molar mass?

Hint 2: What is Raoult's law? How are boiling and freezing points elevated or depressed with solute?

Hint 3: What determines osmotic pressure across a semipermeable membrane (like a cell membrane)?

Cold Beer Anyone?
(problem forty)

Hint 1: What are the components of beer?

Hint 2: What determines the freezing point of the beer?

Hint 3: What changes when you open the can?

Chapter 4 — Chemical, Acid-Base, and Solubility Equilibria

Problem	Title	Topic	Difficulty
41	Profound Perturbations	chemical equilibria	A
42	Food and War	chemical equilibria	A
43	Disappearing Water	chemical equilibria	B
44	Double Dating	chemical equilibria	B
45	Social or Antisocial?	chemical equilibria	C
46	Misbehaved Gases	chemical equilibria	C
47	Pure as Water	acid-base equilibria	A
48	The Status Quo	acid-base equilibria	A
49	Friendly Relations	acid-base equilibria	A
50	A Dash of Salt	acid-base equilibria	B
51	Flex!	acid-base equilibria	B
52	Future Chemistry	acid-base equilibria	B
53	Building Blocks of Life	acid-base equilibria	B
54	Tricky Titration	acid-base equilibria	C
55	Body Buffers	acid-base equilibria	C
56	Bones and Stones	solubility equilibria	A
57	Hard Water	solubility equilibria	B
58	Let It Snow	solubility equilibria	B
59	Ahoy, Captain Silver!	solubility equilibria	C
60	Doctor, Doctor	solubility equilibria	C

Hints 65

Profound Perturbations
(problem forty-one)

Here are a few simple equilibria shorties to illustrate the differences between equilibrium and non-equilibrium conditions. Consider the gas phase equilibrium reaction A <—> B + C. You decide to carry out this reaction in your flask. The equilibrium constant for this reaction is K, and this famous reaction is known to be endothermic.

(i) What is the expression of the equilibrium constant?

(ii) Assume you start with only gas A in your flask. You let the system come to equilibrium. Will the final (equilibrium) pressure be higher or lower than the initial pressure? What is the value of the reaction quotient Q initially and after equilibrium?

(iii) After equilibrium is achieved, are any molecules of A still reacting?

(iv) After equilibrium is achieved, you add more A. Has this perturbation changed the value of K? Has it changed the value of Q? Which direction will the reaction proceed?

(v) You allow the system to re-establish equilibrium. Now you place your flask in the refrigerator (ie, you suddenly lower the temperature). Has this perturbation changed the value of K? Has it changed the value of Q? Which direction will the reaction proceed?

[category: chemical, acid-base, and solubility equilibria]
[topic: chemical equilibria]
[difficulty: A]

Food and War
(problem forty-two)

Dr. Fritz Haber is developing a method for efficiently preparing ammonia in large quantities. He reacts nitrogen gas with hydrogen gas to form gaseous ammonia, which can then be used to make fertilizers. Essentially Dr. Haber is using common gases to make fertilizer, which can be used to grow food. (Unfortunately, during World War I, Germany used this process to make ammonia for explosives, which helped make Germany self-sufficient.)

It is, however, before the outbreak of World War I. You need to feed the starving population and thus want to optimize the Haber process. Note that the process to make ammonia releases heat. What conditions of temperature and pressure would maximize the equilibrium constant? Based on equilibrium considerations only, should you guess high, low, or medium for temperature and pressure in your initial trials?

[Note that the actual process is more complex. Kinetic constraints require the use of a catalyst, and sub-optimal equilibrium conditions have to be used in order to achieve a reasonable rate of reaction. For the purposes of this problem, though, assume that the reaction achieves equilibrium.]

[category: chemical, acid-base, and solubility equilibria]
[topic: chemical equilibria]
[difficulty: A]

Disappearing Water
(problem forty-three)

You walk into your kitchen, fill a glass partially full with water, and leave it out on your kitchen table. Two days later, you return only to find that your glass is dry. If you leave a glass of water open to the air for a day, why does it all evaporate? Is the liquid-vapor system ever in equilibrium? Would the water evaporate more or less quickly on a humid day (assuming, of course, that you leave a window open and do not have air conditioning)?

Consider now the same glass of water, but this time you cover the top with saran wrap (also known as "cling film"). Two days later, you return to find your glass still full of water. Why did the water not evaporate? Is the liquid-vapor system in equilibrium?

In both cases, describe how the reaction quotients and equilibrium constants are related to the vapor pressure of water at room temperature. For a system in equilibrium, can there be any macroscopic changes? For a system in equilibrium, can there be any microscopic changes?

[category: chemical, acid-base, and solubility equilibria]
[topic: chemical equilibria]
[difficulty: B]

Double Dating
(problem forty-four)

You are studying the equilibrium $P_{4(g)} \longleftrightarrow 2P_{2(g)}$. In a previously evacuated sealed vessel, you place a quantity of diatomic phosphorus gas. What is the value of the reaction quotient Q? In which direction will the reaction proceed in order to achieve equilibrium? For this reaction, which is greater, Kp or Kc, or are they equal?

You allow the system to come to equilibrium, and then decide to tamper with it. In which direction would the equilibrium shift if you add $P_{4(g)}$? Increase the pressure of the system (holding temperature constant)? Increase the volume of the system (holding temperature constant)? Increase the temperature of the system (holding pressure constant)?

[category: chemical, acid-base, and solubility equilibria]
[topic: chemical equilibria]
[difficulty: B]

Social or Antisocial?
(problem forty-five)

You have a flask filled with phosphorus pentachloride (PCl_5) at a high enough temperature such that it is entirely in gaseous form. You remember from your college chemistry course that it can decompose into chlorine gas (Cl_2) and phosphorus trichloride gas (PCl_3) in an equilibrium reaction (see illustration on next page). You start with one atm of phosphorus pentachloride in your flask. After equilibrium is achieved, you notice that the new pressure is fifty percent higher than the starting pressure.

You are having lunch with a friend, who is also studying the same reaction at the same temperature. She has combined one atm of each of the three gases in a flask, and is now wondering what will happen. Do you predict that the pressure in her flask will go up, go down, or remain the same? Do you predict that the molecules will be generally "social" (and associate) or "antisocial" (and decompose)? (In other words, will the reaction move towards the left or towards the right)?

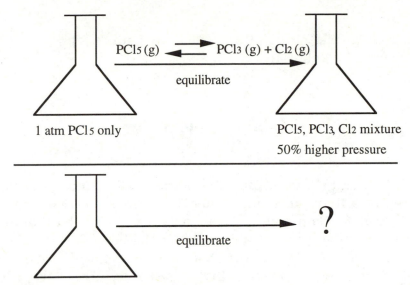

PCl5 (g) ⇌ PCl3 (g) + Cl2 (g)

equilibrate

1 atm PCl5 only

PCl5, PCl3, Cl2 mixture
50% higher pressure

equilibrate

?

1 atm PCl5, 1 atm PCl3, 1 atm Cl2

[category: chemical, acid-base, and solubility equilibria]
[topic: chemical equilibria]
[difficulty: C]

Misbehaved Gases
(problem forty-six)

You place a specific amount of nitrogen dioxide in a previously evacuated sealed vessel of a fixed volume. According to the ideal gas law, the pressure should be directly related to temperature. You find that at high temperatures, a graph of pressure versus temperature is almost exactly linear. However, at low temperatures, the pressure drops much more quickly with temperature than expected. You also find that the gas is more nonideal when you use a higher initial pressure of nitrogen dioxide. What is happening at low and high temperatures? Explain using equilibrium arguments.

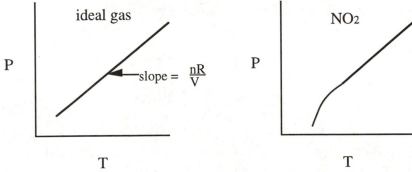

[category: chemical, acid-base, and solubility equilibria]
[topic: chemical equilibria]
[difficulty: C]

Pure as Water
(problem forty-seven)

You have purchased purified bottled water since you know that tap water is full of salts and other impurities. This ultra- super-purified water claims to have only a negligible amount of contaminants that should not make it acidic or basic. You pour yourself a glass and decide to test its pH, expecting it to be precisely neutral. Surprisingly, you find that it is slightly acidic! Why? Think of what happens to gas and liquid molecules at the air-water interface.

[category: chemical, acid-base, and solubility equilibria]
[topic: acid-base equilibria]
[difficulty: A]

The Status Quo
(problem forty-eight)

It is essential to keep a relatively constant pH level in your blood. This is maintained by a number of buffers, such as the carbonic acid/bicarbonate system. Write out this acid-base reaction, and state qualitatively how this buffer solution resists changes in pH. Given the Ka of carbonic acid, how can you determine what pH would make blood the most effective buffer?
Why *can't* you make an effective nitric acid/nitrate buffer?

[category: chemical, acid-base, and solubility equilibria]
[topic: acid-base equilibria]
[difficulty: A]

Friendly Relations
(problem forty-nine)

Chemical equilibria involve complex "sociological" behaviors. Every quantity has a relationship to every other quantity. For example, the Ka of a weak acid is related to the Kb of its conjugate base. Write out the reaction of a general acid HA with water and the reaction of its conjugate base A⁻ with water. If these equilibrium reactions are added together, what is the overall reaction? What is its overall equilibrium constant (in terms of Ka and Kb)? What is its value at room temperature? This is the constant for the "autoionization of water".

Another relationship is between Ka of a weak acid and the pH of a solution. Assume that you begin with a one molar concentration of a monoprotic acid and that only a small fraction of your weak acid or base reacts with water (i.e., Ka is small). What is the pH of the solution in terms of Ka?

[category: chemical, acid-base, and solubility equilibria]
[topic: acid-base equilibria]
[difficulty: A]

A Dash of Salt
(problem fifty)

You're on the spot, in the limelight, poised to win the grand prize... All you have to do is figure out the puzzle in one minute or less. Ready? You have on one side of a table three solutions each containing a different cation (NH_4^+, Na^+, or H^+) with non-reactive anions, and these solutions are labeled in an unknown order A, B, and C. On the other side of the table are two solutions containing different anions (Cl^- or OH^-) with non-reactive cations, and these solutions are labeled Y and Z. All solutions are one molar concentration. You know that an equal mixture of B and Y is acidic, B and Z is basic, and A and Z is neutral. Identify all five unknowns to win fame, riches, and glory.

	Y	Z
A		neutral
B	acidic	basic
C		

[category: chemical, acid-base, and solubility equilibria]
[topic: acid-base equilibria]
[difficulty: B]

Flex!
(problem fifty-one)

The strength of an acid is dependent on the position of the equilibrium HA <—> H$^+$ + A$^-$. An equilibrium favoring the products of this reaction would imply a stronger acid. Compare the aqueous dissociation of the acids HF and HCl. Which one is a weak acid and which is a strong acid? Which one would have a higher dissociation equilibrium constant? Assume you have some reagent that chelates and removes fluoride and chloride ions from solution. If you add a large amount of this reagent to a 1 M solution of HF, would you expect the pH of the solution to change significantly? If you add a large amount of this reagent to a 1 M solution of HCl, would you expect the pH of the solution to change significantly? Consider this in terms of Le Chatelier's Principle.

[category: chemical, acid-base, and solubility equilibria]
[topic: acid-base equilibria]
[difficulty: B]

Future Chemistry
(problem fifty-two)

Now you are looking into a crystal ball, seeing the future... You are taking organic chemistry. You are struggling, wishing you had paid more attention to your general chemistry course. An overarching principle throughout organic chemistry is simple Lewis acid-base reactions.

Consider the simplest acid-base reaction: H$_3$O$^+$ + OH$^-$ <—> 2H$_2$O. Which are the Bronsted acids and Bronsted bases? Identify the acid/base conjugates and note the "proton transfer". How can these species also be considered Lewis acids and bases? Where is the "electron transfer"?

Consider now another acid-base reaction: NH$_3$ + BF$_3$ <—> NH$_3$BF$_3$. Are any of these species Bronsted acids or bases? Which is the Lewis acid and Lewis base? Show the electron transfer in forming the product using orbital diagrams.

Knowing that Lewis bases ("nucleophiles") can attack and bond to the empty orbitals in Lewis acids ("electrophiles"), let us extend this acid-base concept to a typical reaction that you will definitely study if you take an organic chemistry course (see illustration):

$$CH_3-CO-CH_3 + OH^- <—> [CH_3-CO(OH)-CH_3]^-$$
$$[CH_3-CO(OH)-CH_3]^- + H^+ <—> CH_3-C(OH)_2-CH_3$$

This two-step reaction is known as "hydrating a carbonyl" (see illustration). A carbonyl is a structure that has an oxygen double bonded to a carbon, and the hydration reaction adds a water molecule in two steps (OH$^-$ first, then H$^+$). Write the resonance structure of the carbonyl that is important for this reaction. It has formal charge separation and makes the oxygen nucleophilic and the carbon electrophilic. Using this structure, demonstrate how it acts as a Lewis acid in the first step, showing how it acts as an electron acceptor. Then demonstrate how it acts as a Lewis acid in the second step, showing how it acts as an electron donor.

[category: chemical, acid-base, and solubility equilibria]
[topic: acid-base equilibria]
[difficulty: B]

Building Blocks of Life
(problem fifty-three)

Four classes of biological molecules that are essential to life are proteins, carbohydrates, lipids, and nucleic acids. Each of these groups can be described as structures consisting of repeating subunits (see illustrations on next page).

Proteins are polymers consisting of a linear array of amino acids, each with the structure H_2N-CHR-COOH, where R represents a variable side chain. Consider this general structure for a free amino acid. Study in particular the amine and carboxyl groups of the molecule— and note if they are acidic or basic groups. Can the amino acid molecule act readily as an acid? Can it act readily as a base? What would you predict for its structure at rather acidic, basic, or neutral pH's?

Carbohydrates are polymers with repeating subunits of sugar residues, which often have a five- or six-membered ring structure (see illustration on next page). Lipids are composed of several fatty acids that have the structure R-COOH, where R is a hydrocarbon chain. Nucleic acids are polymers of nucleotides, each of which has the structure H_2PO_4-sugar-purine/pyrimidine. Are these subunits (as in the forms depicted) acids, bases, or neither? If they are acids or bases, write out the chemical reaction involved.

For the molecules that you identified as acids or bases, are they "weak" or "strong"?

Four Classes of Biological Molecules

Macromolecule	Subunits
PROTEINS	Amino Acids
CARBOHYDRATES	Simple Sugars
LIPIDS	Fatty Acids
NUCLEIC ACIDS	Nucleotides

amino acid:

amine group   carbonyl group

sugar:

fatty acid:

nucleotide:

HO–P–O– sugar– purine/pyrimidine

[category: chemical, acid-base, and solubility equilibria]
[topic: acid-base equilibria]
[difficulty: B]

Tricky Titration
(problem fifty-four)

You are performing a number of titration experiments. You first perform a simple one- You place 10 mL of 0.1 M HCl in a flask and add 0.1 M NaOH gradually, and you obtain the graph in panel A. After you add 10 mL of the base, you are at the "equivalence point"— what would you expect to be the pH of the solution at this point? Why?

Next, you decide to study acetic acid. You place 10 mL of a 0.1 M acetic acid solution in a flask and add 0.1 M NaOH gradually, and you obtain the graph in panel B. Why is the initial part of the graph flat?

After you add 10 mL of the base, you are at the "equivalence point"—would you expect the pH of the solution at this point to be greater, less than, or equal to 7? Why?

      Finally, you do a tricky titration. You place 10 mL of a 0.1 M phosphoric acid solution in a flask and add 0.1 M NaOH as above, and you obtain the graph in panel C. Why do you get such an odd shape? Write down the relevant chemical reactions. What are the predominant species present at points I, II, III, IV, and V? You notice that there appear to be only two equivalence points, at 10 mL and 20 mL of base. This puzzles you, since you know phosphoric acid is a *tri*protic acid. Why do you not see a third equivalence point?

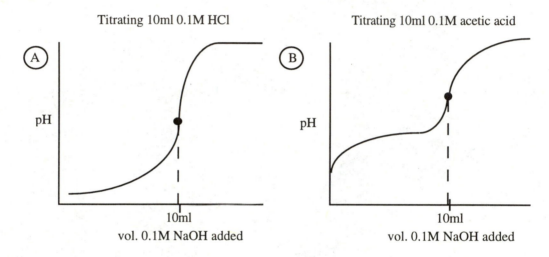

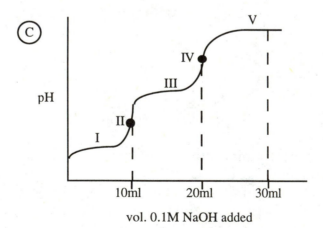

[category: chemical, acid-base, and solubility equilibria]
[topic: acid-base equilibria]
[difficulty: C]

Body Buffers
(problem fifty-five)

The human body includes buffers so that a person's blood pH can be maintained within a narrow margin. The most important physiological buffer involves carbonic acid (H_2CO_3) and bicarbonate (HCO_3^-). The carbonic acid concentration is determined by the partial pressure of carbon dioxide gas (pCO_2) in the lungs, which is also dissolved in the blood to form carbonic acid:

$$\text{lung } CO_{2(g)} <\!\!-\!\!> \text{blood } CO_{2(g)} <\!\!-\!\!> \text{blood } CO_{2(aq)}$$
$$CO_{2(aq)} + H_2O <\!\!-\!\!> H_2CO_{3(aq)} + H_2O <\!\!-\!\!> HCO_3^-{}_{(aq)} + H_3O^+{}_{(aq)}$$

Carbon dioxide is produced by all living cells and is removed from the body by the lungs during exhalation. Normally small changes in the rate of breathing have little effect on a person's pH, but breathing very slowly (hypoventilation) or very quickly (hyperventilation) cause changes marked changes in pCO_2 levels and thus also pH. Furthermore, addition or depletion of other physiological acids and bases usually has little effect on pH, but in severe cases this can have significant effects. A person's blood pH is generally between 7.36 and 7.44. If it rises above this range, the medical condition is called "alkalosis", and if it drops below this range then it is known as "acidosis". Use this background and your knowledge of buffers to analyze the following cases:

Case 1. A patient is rushed to the emergency room because of a severe asthma attack that has made it impossible to take deep breaths.

Case 2. A patient goes to the top of a mountain and starts to hyperventilate due to the thin air.

Case 3. A diabetic cannot metabolize sugars properly and has to metabolize fats instead, which produces large amounts of acidic by-products (known as ketoacids).

Case 4. A patient has a stomach flu and vomits repeatedly, thus losing large amounts of stomach acid (specifically HCl).

For each case, you do a quick blood test to measure pH, pCO_2, and bicarbonate levels. How would you expect these variables to compare with normal values in each case (ie which ones are high and which are low)? Which of these variables (pH, pCO_2, or bicarbonate) would have the smallest percentage deviation from normal and why?

[category: chemical, acid-base, and solubility equilibria]
[topic: acid-base equilibria]
[difficulty: C]

Bones and Stones
(problem fifty-six)

In the body there are many mechanisms of regulating the levels of calcium in the blood. Bone is a reservoir of calcium that can be partially solubilized when more aqueous calcium is needed. When calcium levels in the blood drop, the parathyroid glands (located in your neck) release a hormone called PTH (the not-so-imaginative name, parathyroid hormone). One of the major effects of this hormone is to cause solidified calcium in the bones to be absorbed into the bloodstream as free aqueous ions. It is known that calcium ions and phosphate ions usually circulate in the blood near their saturation points. The precipitation of calcium phosphate crystals is a very relevant chemical reaction since they can become lodged in the tubules of the kidney causing a painful but rather common condition called a kidney stone.

You are a chemist who is called upon to aid your friend, a confused doctor. Your friend tells you that his patient has been diagnosed with hyperparathyroidism — this is a condition in which the parathyroid glands secrete an inappropriately high amount of PTH. Your friend asks you to please explain in chemical terms how this disorder can lead to (1) osteoporosis (weakening of the bones) and (2) kidney stones. Write out the relevant phase or solubility reactions.

[category: chemical, acid-base, and solubility equilibria]
[topic: solubility equilibria]
[difficulty: A]

Hard Water
(problem fifty-seven)

Hard water is often caused by small quantities of salts, especially calcium salts, that are dissolved in rainwater and rivers. A typical concentration of calcium sulfate in hard water is a few milligrams per liter. In order to soften the water, you hope to add some compound that will precipitate the calcium ions from solution. You know that the solubility product constant of calcium phosphate is very small and that of calcium carbonate is much larger. You have four reagents at your disposal: (1) concentrated HCl; (2) concentrated NaOH; (3) a buffer system containing sodium phosphate monobasic (NaH_2PO_4) and dibasic (Na_2HPO_4); and (4) a buffer system containing sodium carbonate

(Na_2CO_3) and sodium bicarbonate ($NaHCO_3$). What is the best procedure for softening the water?

[category: chemical, acid-base, and solubility equilibria]
[topic: solubility equilibria]
[difficulty: B]

Let It Snow
(problem fifty-eight)

You have a huge beaker of salty water. Three salts (sodium chloride, sodium iodide, and sodium bromide) are all at approximately the same concentration. You hook up the beaker to a magnetic stirring system to insure that it is always well mixed, and then you very gradually add a dilute solution of silver nitrate. You see a precipitate forming, and you turn the stirrer off and allow the solid to settle at the bottom of the beaker. You remove the precipitate and label it "Snowfall #1". You then add more silver nitrate and obtain more precipitate, which you label "Snowfall #2", and then repeat the procedure once more to obtain "Snowfall #3". You then add a lot more silver nitrate and nothing happens. What are the solids in each snowfall?
　　Note that the Ksp(AgCl) > Ksp(AgBr) > Ksp(AgI).

[category: chemical, acid-base, and solubility equilibria]
[topic: solubility equilibria]
[difficulty: B]

Ahoy, Captain Silver!
(problem fifty-nine)

You are investigating the properties of silver ions in solution. You place silver nitrate in a liter of water and notice that a large amount dissolves to form a solution. Then you add a teaspoon of table salt (sodium chloride) and note what happens in your lab booklet. You then add a teaspoon more of water to see what would happen. Finally, you add a teaspoon of pure ammonia and note what happens. Describe qualitatively what would happen at each step.
　　The Ksp for silver chloride is very small, the Ksp for silver nitrate is relatively large, and the Keq for the formation of the

$Ag(NH_3)_2^+$ complex ion is very large (many orders of magnitude difference).

$$AgNO_3 \text{ (aq)} \longrightarrow Ag^+ \text{ (aq)} + NO_3^- \text{ (aq)}$$

$$Ag^+ \text{ (aq)} + Cl^- \text{ (aq)} \rightleftharpoons AgCl \text{ (s)}$$

$$Ag^+ \text{ (aq)} + 2\,NH_3 \text{ (aq)} \rightleftharpoons Ag(NH_3)_2^+$$

[category: chemical, acid-base, and solubility equilibria]
[topic: solubility equilibria]
[difficulty: C]

Doctor, Doctor!
(problem sixty)

Barium sulfate is used in medicine for radiographic investigations of the gastrointestinal tract, because it is opaque to radiation and has a low Ksp. Solvated barium cations are normally rather toxic. The low Ksp implies that not very much barium sulfate dissociates into the harmful barium ions.

You are assisting a physician in the hospital, and the patient is suffering from some intestinal disorder. The doctor concludes that a radiographic visualization is necessary and gives the patient a large oral dose of barium sulfate. The patient begins to have an unusual allergic reaction to the small quantity of free barium ions that *are* dissolved, severe enough to warrant immediate attention. Unfortunately, the doctor has forgotten his chemistry, and there is no time for him to look up anything in his books. He turns to you.

What can you give the patient to remove rapidly the free barium ions from his digestive system? (Any common salt would be readily available.) For the most effective results, should you tell the patient to drink a cup of water, grapefruit juice, nitric acid, water with a dissolved antacid tablet, or lye?

[category: chemical, acid-base, and solubility equilibria]
[topic: solubility equilibria]
[difficulty: C]

Chemical, Acid-Base, and Solubility Equilibria — HINTS

Profound Perturbations
(problem forty-one)

Hint 1: How is the law of mass action used to write out the equilibrium constant? What is the relationship between K and Q?

Hint 2: What is Le Chatelier's Principle? How is it applied to changing the concentration or temperature of the system?

Hint 3: Is K dependent on temperature?

Food and War
(problem forty-two)

Hint 1: What is the balanced equilibria reaction for the formation of ammonia from nitrogen and hydrogen?

Hint 2: In which direction should you shift the equilibrium in order to maximize yield?

Hint 3: State the general form of Le Chatelier's Principle. How does it apply to temperature and pressure?

Disappearing Water
(problem forty-three)

Hint 1: What is the liquid-vapor equilibrium reaction? What is an expression for the equilibrium constant?

Hint 2: Is a system in equilibrium static, or is the system dynamic with balanced opposing reactions?

Hint 3: Why does sealing the glass make the difference between whether the water evaporates or not?

Double Dating
(problem forty-four)

Hint 1: How does the relationship between Kp and Kc reflect the ideal gas equation?

Hint 2: What is Le Chatelier's Principle?

Hint 3: Is this reaction endothermic or exothermic?

Social or Antisocial?
(problem forty-five)

Hint 1: What is the equilibrium constant (Kp) of the reaction?

Hint 2: In your experiment (first paragraph), what conclusions can be drawn about the equilibrium constant of the chemical reaction?

Hint 3: In your friend's experiment (second paragraph), what is the value of Q? Which direction will the reaction proceed?

Misbehaved Gases
(problem forty-six)

Hint 1: What reaction of nitrogen dioxide is promoted by low temperatures?

Hint 2: How does this make the gas very nonideal? What assumption of the kinetic theory of ideal gases is violated?

Hint 3: Why should the nonideality become pronounced at higher pressures?

Pure as Water
(problem forty-seven)

Hint 1: What happens to gas and liquid molecules at the phase boundary? What is Henry's law?

Hint 2: What are the major components of air?

Hint 3: Which molecule could react as a weak acid?

The Status Quo
(problem forty-eight)

Hint 1: If a small amount of a strong acid is added to a buffered solution, with what does it react? How about if a strong base is added? What if a strong acid or base were added to pure water?

Hint 2: Is the reaction involving carbonic acid and bicarbonate an equilibrium or a reaction that proceeds to completion? How about the reaction between nitric acid and nitrate?

Hint 3: What is the Henderson-Hasselbalch equation? How does this tell you the optimal pH for buffers?

Friendly Relations
(problem forty-nine)

Hint 1: When reactions are added together, how are their equilibria constants treated?

Hint 2: What are the concentrations of hydroxide and hydronium ions in a neutral aqueous solution? How much water is autodissociated?

Hint 3: How can you set up an equilibria expressions knowing only initial conditions?

A Dash of Salt
(problem fifty)

Hint 1: What is the fundamental definition of pH/acidity/basicity in terms of hydrogen ion concentration? Can other species (hydroxide, a metal, etc.) affect the hydrogen ion concentration indirectly?

Hint 2: What is the relationship between the reactivity of an ion in solution and its conjugate acid/base? What are the conjugate acids/bases to the ions above?

Hint 3: What is the definition of a strong and weak acid/base? How does the relative strength of two species in solution determine the overall acidity/basicity?

Flex!
(problem fifty-one)

Hint 1: What is the relationship between equilibrium constant and the extent of the reaction?

Hint 2: How does the extent to which the reaction proceeds determine the pH of the solution?

Hint 3: What is Le Chatelier's Principle? How would the equilibrium respond to removing the fluoride or chloride from solution?

Future Chemistry
(problem fifty-two)

Hint 1: What are the definitions of a Bronsted acid and base (in terms of proton donating and accepting)?

Hint 2: What are the definitions of a Lewis acid and base (in terms of electron pair donating and accepting)?

Hint 3: How is the hydration of a carbonyl merely two Lewis acid-base reactions?

Building Blocks of Life
(problem fifty-three)

Hint 1: Is ammonia an acid or base? Is acetic acid an acid or base? At neutral pH, in what form do each of these exist?

Hint 2: How do the illustrated structures suggest acidic or basic groups?

Hint 3: If biochemical systems had strong acids and bases, could their chemistry be controlled? How about weak acids and bases?

Tricky Titration
(problem fifty-four)

Hint 1: What is the reaction between a strong acid and base? When you add equimolar amounts of this acid and base do you have a neutral solution?

Hint 2: What happens when a strong base reacts with a weak acid? When is a "buffer" formed? When you add equimolar amounts of this acid and base do you have a neutral solution?

Hint 3: Why do polyprotic acids give rise to multiple equivalence points? What are the relative magnitudes of the Ka's of phosphoric acid? How is this related to the absence of the predicted "third" equivalence point?

Body Buffers
(problem fifty-five)

Hint 1: What is the function of buffers? What happens to the carbonic acid/bicarbonate equilibrium if you increase or decrease the carbonic acid levels? How about if you add a small amount of strong acid or base?

Hint 2: What is the Henderson-Hasselbach equation?

Hint 3: What is the relationship between $pCO_{2(g)}$ and the rate of breathing (ie, breathing too slow or breathing too fast)?

Bones and Stones
(problem fifty-six)

Hint 1: What is the solubility equilibria reaction between calcium and phosphate?

Hint 2: Does the calcium term in the above reaction include calcium stored in bone or calcium in the blood or both?

Hint 3: What is the phase equilibria reaction between calcium in bone and calcium in the blood?

Hard Water
(problem fifty-seven)

Hint 1: What are the definitions of solubility product constants? What does it mean for them to be very small or very large?

Hint 2: Which compound do you want to make to remove calcium ions from solution? Write a balanced chemical equation.

Hint 3: At what pH would the reaction be most effective?

Let It Snow
(problem fifty-eight)

Hint 1: In what form are the salts in the initial beaker of salty water?

Hint 2: Write out the solubility reactions for the three silver salts. Which one is least soluble? Which one is precipitated first?

Hint 3: Why is it necessary to know that addition of more silver nitrate at the end yields no more precipitates?

Ahoy, Captain Silver!
(problem fifty-nine)

Hint 1: What are the chemical equations of all the processes? Where does the equilibrium of each lie?

Hint 2: What is LeChatelier's Principle? How does it relate to this problem?

Hint 3: Is AgCl very soluble? How about $Ag(NH_3)_2^+$?

Doctor, Doctor!
(problem sixty)

Hint 1: What is the solubility reaction for barium sulfate?

Hint 2: What is Le Chatelier's principle, and how can this be applied to a solubility problem? This is called the common ion effect. Which direction do you wish to shift the equilibrium?

Hint 3: At which pH would your treatment be most effective?

Chapter 5 — Kinetics

Beyond All Doubt
(problem sixty-one)

You just came up with an ingenious third order mechanism for the formation of nitrogen dioxide from NO and O_2, and decide to test it experimentally. You walk into your lab and measure the rate of a standard reaction. You find that when the concentration of NO is doubled, the rate is quadrupled, and when the concentration of both species are doubled, the rate is octupled. Write the rate law for this reaction. Does this prove that your mechanism is correct?

[category: kinetics]
[topic: reaction rates]
[difficulty: A]

Wining and Dining
(problem sixty-two)

You have before you two glasses of Amontillado wine — one made last year and one made many years ago. You have claimed to be an expert wine connoisseur and able to "taste the age" of the wine. If you can guess which one is older and its age, then you receive a cask of this precious wine.

Unfortunately, you have a cold and cannot taste any difference between the two. Luckily, you have a handy pocket instrument capable of detecting tritium decay. Tritium undergoes radioactive decay with a half-life of slightly over a decade. The glass on the left has sixty-four times fewer tritium counts than the glass on the right. What are the approximate dates of both vintages? What assumptions did you have to make in order to compare their tritium counts? Is this an accurate assumption in light of the events of the post-World War II world?

tritium counts: lower   higher

[category: kinetics]
[topic: reaction rates]
[difficulty: A]

Get Well Soon
(problem sixty-three)

The thermal denaturation (deactivation) of a virus is a first order process. You measure the rate constant at a number of temperatures and plot graphically the natural log of the rate constant versus the reciprocal of the absolute temperature. Your graph is linear and has a negative slope and a positive y-intercept (see illustration).

To what quantities do the slope and y-intercept of this graph correspond? Does this process have a positive or negative activation energy? Does this mean that the deactivation of the virus is faster or slower when a person has a fever?

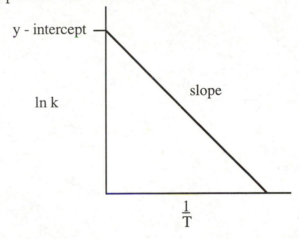

[category: kinetics]
[topic: reaction rates]
[difficulty: A]

Up and Down
(problem sixty-four)

Your job is to determine the rate law for the famous aqueous reaction X—>Y. You perform some experiments and show that doubling the concentration of X doubles the initial reaction rate, and the amount of Y present has no effect on the initial reaction rate.

You then decide to make some graphs to illustrate the kinetics of the reaction. In the first experiment, you vary the initial concentration of X and measure the initial rate. You plot rate vs. [X], as depicted in panel A. You then do a second experiment, in which you use an initial quantity of X and follow the disappearance of the reactant over time. You plot ln [X] versus time, as depicted in panel B.

In your two graphs, one has an upward slope and one has a downward slope. How are these two slopes related?

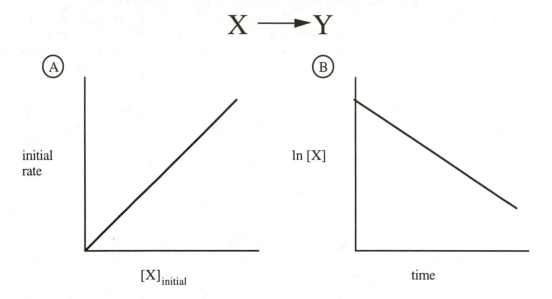

[category: kinetics]
[topic: reaction rates]
[difficulty: B]

Easy as One, Two, Three
(problem sixty-five)

What is the most likely kinetic order for a simple unimolecular decay reaction (that does not depend on solvent effects)? How about for a simple bimolecular interaction, assuming that the reaction involves a single bimolecular collision?

You are studying a chemical reaction A—>B, which involves only A as the reactant. The product B is easily assayed, and you can measure the initial rate of the reaction by measuring the formation of B. You perform two kinetic experiments involving the same initial number of A molecules. The reactions are carried out either (i) with concentrated reactant in a small volume or (ii) with dilute reactant in a large volume. Which experiment would yield a faster initial formation of B molecules if the reaction were first order? How about if the reaction were second order?

[category: kinetics]
[topic: reaction rates]
[difficulty: B]

Science and Medicine
(problem sixty-six)

A large number of chemicals are used both in the laboratory as investigational agents as well as in hospitals as pharmaceuticals. An example in radioactive iodine. It is an isotope that is used frequently in biochemistry labs to label proteins, and it is used in clinical settings to diagnose a number of thyroid diseases such as hyperthyroidism. It emits high energy gamma radiation and thus makes it easy to visualize and follow as a "tracer", either during biochemical reactions in the lab or in a patient body.

In the lab, there is usually only one relevant half-life, namely the time it takes for half of the iodine to decompose. In a patient, however, there are two important half-lives (both first order)— the time it takes for half the iodine to decompose ($t_{1/2,\,1}$) and the time it takes for half the iodine to be excreted from the body via the kidneys ($t_{1/2,\,2}$). Would the *overall* half-life in a patient, namely the time for half the iodine to lose effectiveness, be more or less than in the lab? What would this half life be (in terms of $t_{1/2,\,1}$ and $t_{1/2,\,2}$)?

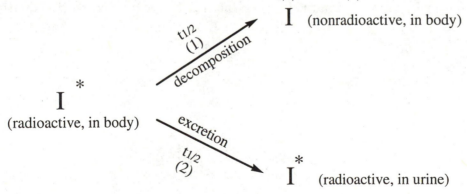

[category: kinetics]
[topic: reaction rates]
[difficulty: B]

Drug Overdose
(problem sixty-seven)

The branch of kinetics that deals with the biological actions of drugs is known as pharmacokinetics. Most drugs exist in two forms: free in the blood and bound by tissues. The drug that is free in the blood is able to perform its specific pharmacological actions as well as to exert its unintended side effects. In contrast, the drug bound to and sequestered by the body tissues (generally fat or muscle tissue) acts as a "reservoir" of large quantities that exert no effects until released into the bloodstream. Thus a good model for the kinetics of action of a drug D is as follows:

D (blood, active) <—> D (tissue bound, inactive)

A cancer patient was to receive a certain dosage of a potent chemotherapeutic drug, but this dosage was to be delivered gradually via an intravenous drip over the course of several hours. The nurse, however, gave the entire dosage as a single bolus injection. Would you characterize these two methods of delivering drugs to the blood as instantaneous, zero-th order, or first order kinetics? How about the kinetic order of the drug being distributed to the tissues?

Based on the above kinetic model, would both methods of delivering the drug be equivalent since the same amount of drug is given in both cases? Would one of the methods yield a higher drug concentration in the blood in the short term? How about after equilibrium is achieved? Explain.

[The above senario was a true, widely-publicized situation that occurred in a prestigious U.S. hospital and tragically led to the death of the patient.]

[category: kinetics]
[topic: reaction rates]
[difficulty: C]

Circular Reasoning
(problem sixty-eight)

In many biological systems, a chemical or biochemical reaction exhibits a phenomenon known as a feedback inhibition. In other words, as a product is made, it inhibits the further production of that product. It would be very disadvantageous and wasteful for a cell to keep making a metabolite when it already possesses enough of that substance, and it would be disastrous for it not to make a metabolite when it is needed. The feedback inhibition allows the chemical process to continue until there is enough of the product. This process allows a cell or other living system to maintain control of the concentrations of its important materials.

For example, consider a biochemical reaction that converts threonine to isoleucine. Threonine and isoleucine belong to a class of molecules called "amino acids" and are important in building proteins. This reaction is particularly significant for maintaining a high enough concentration of isoleucine to sustain life. It is catalyzed by an enzyme that is turned off when the concentration of isoleucine are high. This saves the cell much metabolic energy, because there is no need to make isoleucine when the cell already possesses enough.

Consider a simple reaction A —> B. What is a likely rate law if this is a simple enzyme catalyzed reaction (Michaelis-Menten kinetics)? Assume now that the enzyme is at a sufficient concentration that it is not saturated. What rate laws are possible if there is no feedback control? What rate laws are possible if there is feedback inhibition? What rate laws are possible if the opposite situation occurs, that is, the product *increases* its own production? This last case is the opposite of feedback inhibition and is known as feedback stimulation or positive control.

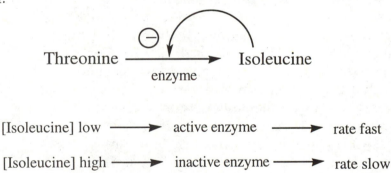

[Isoleucine] low ——→ active enzyme ——→ rate fast

[Isoleucine] high ——→ inactive enzyme ——→ rate slow

Feedback Inhibition   Feedback Stimulation

[category: kinetics]
[topic: reaction rates]
[difficulty: C]

Shapes and Slopes
(problem sixty-nine)

You are studying the isomerization of your reactant, CH_3NC, to the related species CH_3CN. This reaction proceeds spontaneously without the need of any catalysts, other reagents, or solvent molecules. If this isomerization is an elementary reaction step, what would you predict to be the most likely kinetic order of the reaction? What is the rate law for the reaction?

Assuming that your CH_3NC reacts to completion, sketch how the concentration of your reactant depends on time. Also sketch the rate of the reaction versus time. What mathematical form (cubic, exponential, logarithmic, etc.) best describes the general shape of these curves?

[category: kinetics]
[topic: mechanisms]
[difficulty: A]

Iconoclast!
(problem seventy)

The reaction between diatomic hydrogen and diatomic iodine to produce hydrogen iodide is a chemical classic. For nearly a hundred years it was thought to occur by a simple bimolecular reaction. Write out the mechanism for this interpretation of this reaction, and show that it supports the experimentally determined second order rate law.

Another possible mechanism involves three steps as follows:
    (i) $I_2 <\longrightarrow 2I$ (fast equilibrium)
    (ii) $H_2 + I <\longrightarrow H_2I$ (fast equilibrium)
    (iii) $H_2I + I \longrightarrow 2HI$ (slow, rate determining step)

What rate law does this mechanism predict? Is it the same as the first, simpler mechanism?

An experiment was then performed that demonstrated the presence of free iodine atoms as intermediates in the reaction. What is the implication for the simple, traditional mechanism? Does this prove that the new mechanism is the correct one?

[category: kinetics]
[topic: mechanisms]
[difficulty: B]

Steady as She Goes
(problem seventy-one)

The steady-state approximation is a rather general technique in analyzing reaction mechanisms. Analyzing a mechanism by this technique can lead to a rate law that, under the appropriate conditions, can reduce to the rate law that would be predicted if one step were rate determining. For example, consider the following mechanism:

$$A + 2B \longleftrightarrow C + D \qquad \text{(forward rate constant } k_1\text{)}$$
$$\text{(reverse rate constant } k_{-1}\text{)}$$
$$C + E \longrightarrow F \qquad \text{(rate constant } k_2\text{)}$$

What is the overall reaction? What are the intermediate(s)? Write a differential equation defining the rate of the reaction in terms of the concentration of either A or F.

Analyze this mechanism under three conditions: (i) if we know that the first step is a very slow rate determining step; (ii) if we know that the second step is a very slow rate determining step; or (iii) if we make no assumptions on the relative rates of the two steps. For the first two conditions, the rate law can easily be determined with only minimal algebraic manipulations. For the third case, make a steady-state approximation on the reaction intermediate.

Now show that the general rate law derived using the steady-state approximation reduces to the first case when k_1, $k_{-1} << k_2$ (i.e., the first step is slow). Also show that it reduces to the second case when k_1, $k_{-1} >> k_2$ (i.e., the second step is slow).

[category: kinetics]
[topic: mechanisms]
[difficulty: B]

Ways and Means
(problem seventy-two)

You are studying the chemical reaction in which a hydroxide ion displaces the chloride in methyl chloride to form methanol:

$$CH_3Cl + OH^- \longrightarrow CH_3OH + Cl^-$$

You are curious to know exactly how the hydroxide ion displaces the chloride in this simple reaction. You realize that there are two plausible mechanisms, either (i) a one-step mechanism in which the two reactants collide and one ion displaces the other, or (ii) a two step mechanism in which the chloride first dissociates to form a CH_3^+

intermediate, which then combines very rapidly with a hydroxide ion (see illustration).   Experimentally, you find that the reaction rate doubles when you double the concentration of either reactant. Which mechanism is more likely to be correct?

Imagine methyl chloride as a tetrahedral molecule in space. You realize that if a chloride leaves the molecule, then the hydroxide ion must bind to the central carbon either from the opposite side (for mechanism i) or after the chloride has left (for mechanism ii). What, then, is the structure of the transition state of the reaction intermediate based on your choice of mechanism above?

You notice two further effects experimentally. You discover that the rate of the reaction is greatly enhanced if you add a certain enzyme catalyst, and that the reaction is also speeded up if you run it at a higher temperature. With the aid of a reaction coordinate diagram, explain the first observation. With the aid of the Maxwell-Boltzmann velocity distribution curve, explain the most obvious reason for the second phenomenon. Give also another reason for the second phenomenon from your reaction intermediate diagram, keeping in mind that chemical bonds are harmonic oscillators and that temperature increases the frequency of oscillation.

(i)

one-step mechanism

(ii)

two-step mechanism

[category: kinetics]
[topic: mechanisms]
[difficulty: C]

Kinetic Control
(problem seventy-three)

In many systems, whether or not a reaction occurs is determined by kinetic control. For example, a reaction that is thermodynamically spontaneous might be prevented from occuring under most circumstances due to a high activation energy barrier. Only in the presence of a catalyst would the activation energy barrier be lowered, thus permitting the reaction to proceed. A number of different types of molecules can be catalysts, such as biochemical enzymes, surface-adsorbed heavy metals, etc.

Consider the reaction A <—> B, which has a very high activation energy barrier. You place in a vessel equal quantities of A and B, and no reaction occurs. You then place in the vessel a good catalyst- What do you expect to happen if the equilibrium constant for the reaction is greater than one? less than one? equal to one?

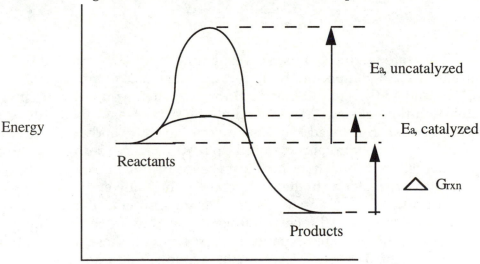

[category: kinetics]
[topic: catalysis]
[difficulty: A]

Powder Power
(problem seventy-four)

There is a fundamental difference between reactions that occur in solution or the gaseous phase and reactions that involve solids or liquids. With aqueous or gaseous systems, each species fills the entire volume and is dispersed evenly, and the rate is related to the pressure or concentration. When a solid or liquid reacts, the situation is different. Consider the reaction of a piece of solid sodium reacting with water. What is this reaction? Would it be more explosive if you use a chunk of sodium or the same mass of finely powdered sodium? Where (physically) is the reaction taking place? Can you express the rate as merely dependent on the concentrations of the reactants?

Analogous questions include: Why do you chop wood before lighting a fire in your fireplace? Why is it better to chew food before swallowing? Why is grain dust so explosive but grain relatively safe?

[category: kinetics]
[topic: catalysis]
[difficulty: B]

Protein Magic
(problem seventy-five)

You are a biochemist investigating the transformation of compound A to compound B: A —> B. You place compound A in a test tube and sadly find out that no B forms. Then you add a drop of a protein solution labeled "Magic" and all the A is transformed into B in a few seconds. Surprised, you assay your mixture and find out that none of your magic protein was used up!! You decide to do a few more experiments. You find that, in the presence of the magic protein, the reaction occurs faster at higher temperatures but slower at *much* higher temperatures. In the absence of the magic protein, the reaction is undetectable at these "higher" temperatures, but it occurs slowly at these "much higher" temperatures. Explain all these observations with the aid of a reaction coordinate diagram and molecular speed distribution curves. At *extremely* high temperatures, would you expect the reaction to be faster with the magic protein or without it?

[category: kinetics]
[topic: catalysis]
[difficulty: C]

Kinetics — HINTS

Beyond All Doubt
(problem sixty-one)

Hint 1: What is the balanced chemical reaction being investigated?

Hint 2: How can the experiment determine the order of the reaction?

Hint 3: What is the strength of the relationship between experimental rate law determination and theoretical mechanisms?

Wining and Dining
(problem sixty-two)

Hint 1: Is radioactive decay first order?

Hint 2: What is the mathematical relation for the time dependence of a radioactive decay process?

Hint 3: How has the proliferation of nuclear technology altered the amount of tritium in the atmosphere?

Get Well Soon
(problem sixty-three)

Hint 1: What is the Arrhenius equation?

Hint 2: How can it be rearranged into a form useful for this problem?

Hint 3: What is the relation between activation energy and temperature dependence of the rate?

Up and Down
(problem sixty-four)

Hint 1: What does the first paragraph say about the kinetic order in the rate law?

Hint 2: In the first plot, what is the slope of the graph?

Hint 3: What is the time-dependence of disappearance of X? What is the slope of this graph?

Easy as One, Two, Three
(problem sixty-five)

Hint 1: What is the difference between a stoichiometric reaction and an elementary reaction step? Is there any difference in the first two examples?

Hint 2: What are the differential equations describing the rate of a first and second order reaction?

Hint 3: What's more important in first and second order reactions, the number of molecules present or their concentration?

Science and Medicine
(problem sixty-six)

Hint 1: Are half-lives additive?

Hint 2: What is the relationship between half-life and the rate constant?

Hint 3: What are all the relevant reactions involved?

Drug Overdose
(problem sixty-seven)

Hint 1: Which component of the system does the drug enter, the blood or the tissues?

Hint 2: Imagine physically the two methods of administering the drug. How do the two methods of giving the drugs differ kinetically?

Hint 3: Is there any difference in the equilibrium situation after the drug has distributed itself between the blood and tissues?

Circular Reasoning
(problem sixty-eight)

Hint 1: What are Michaelis-Menten kinetics? How does enzymatic saturation affect reaction rate?

Hint 2: What is a general rate law (using variable exponents) that can account for all three cases?

Hint 3: What restrictions are there on the exponents of the rate law when there is no feedback control, negative feedback (inhibition), or positive feedback (stimulation)?

Shapes and Slopes
(problem sixty-nine)

Hint 1: Write out the reaction. What is its molecularity?

Hint 2: What is the mathematical equation that describes the time dependence of concentration for this type of reaction?

Hint 3: As the reaction proceeds, the reactant concentration goes down. How does this affect the rate of the reaction?

Iconoclast!
(problem seventy)

Hint 1: How many steps is the simple bimolecular reaction?

Hint 2: What does a slow, rate determining step imply about the rate of the reaction? How can the fast pre-equilibria be used to eliminate the intermediates from the rate law?

Hint 3: Can a mechanism be absolutely disproven? Can a mechanism be absolutely proven?

Steady as She Goes
(problem seventy-one)

Hint 1: How does knowing the slow step of the reaction yield a very accurate approximation of the rate law?

Hint 2: What is the definition of the steady-state? Answer qualitatively as well as mathematically.

Hint 3: Why is it reasonable to expect the steady-state rate law to reduce to these cases under certain specific conditions?

Ways and Means
(problem seventy-two)

Hint 1: What is the rate law for the reaction? What is its order and molecularity?

Hint 2: How can knowing that molecules take up space and that electron clouds tend to repel each other help explain the structure of the reaction intermediate?

Hint 3: The rate of the reaction depends on how frequently the reactants collide, whether they possess enough energy to react, and whether they possess the right orientation to react. Which of these variables are affected by adding a catalyst and increasing the temperature?

Kinetic Control
(problem seventy-three)

Hint 1: How does a catalyst change the equilibrium of a reaction?

Hint 2: How does a catalyst change the kinetics of a reaction?

KINETICS filler

Hint 3: What is the relationship between activation energy and rate?

Powder Power
(problem seventy-four)

Hint 1: What gas is produced when an alkali metal reacts with water?

Hint 2: Where is the contact between a solid and liquid reactant? How is this related to the rate?

Hint 3: How can the physical state of a solid or liquid reactant be related to an "effective concentration" that takes part in the rate law?

Protein Magic
(problem seventy-five)

Hint 1: What type of substance is the magic protein? How does it affect the kinetics and thermodynamics of the reaction? Why?

Hint 2: What happens to the molecular energies at high temperatures? Would this generally increase or decrease the rate of reaction?

Hint 3: How can you explain the increase and then decrease in rate with increasing temperature using the magic protein? Think about two competing effects.

Chapter 6 — Thermodynamics and Electrochemistry

Problem	Title	Topic	Difficulty
76	Stately Paths	q, w, E	A
77	Cold Sweats	q, w, E	A
78	Hot Hot Hot	q, w, E	B
79	First Things First	q, w, E	B
80	What's Hot?What's Not?	q, w, E	B
81	Cool It Down!	q, w, E	C
82	Mountain Madness	q, w, E	C
83	Diamonds Are Forever	H, S, G	A
84	Doing the Impossible	H, S, G	A
85	Everything in Disarray	H, S, G	A
86	Molecular Machines	H, S, G	A
87	Cages of Water	H, S, G	B
88	Cracking the Egg	H, S, G	B
89	Chemical Glue	H, S, G	B
90	Sherlock	H, S, G	C
91	Yin-Yang	H, S, G	C
92	Turn Up the Heat	H, S, G	C
93	Really, Really Cold	H, S, G	C
94	Speeding Bullets	electrochemistry	A
95	Iron Breathing	electrochemistry	A
96	Priceless Photos	electrochemistry	B
97	Silly Cells	electrochemistry	B
98	Varying Volts	electrochemistry	B
99	Flowing Ions	electrochemistry	C
100	Waterfall of Energy	electrochemistry	C

Stately Paths
(problem seventy-six)

You and two friends are studying the thermodynamic properties of a certain ideal gas. You all are experts on the ideal gas law, and know that as you increase the pressure of an ideal gas the volume goes down (PV=constant if temperature doesn't change). You all decide to study the compression of your gas from V_{hi} to V_{lo} due to increasing the pressure from P_{lo} to P_{hi} and ensuring that the temperatures of the initial and final states are the same. However, you and your friends go about this differently (see illustration on next page).

- You first cool the gas at constant pressure to lower the volume, and then you heat the gas at constant volume to raise the pressure (panel A).

- One friend heats the gas at constant volume to raise the pressure, and then cools the gas at constant pressure to lower the volume (panel B).

- Your other friend keeps temperature constant the whole time and simultaneously lowers the volume and increases the pressure (panel C).

Is the overall energy change the same in all three cases? Are the overall heat and work the same in all three cases? Of the quantities P, V, q, w, and E, which are "state" and which are "path" functions?

State Functions vs. Path Functions

Pressure (P)
Volume (V)
Heat (q)
Work (w)
Energy (E)
?

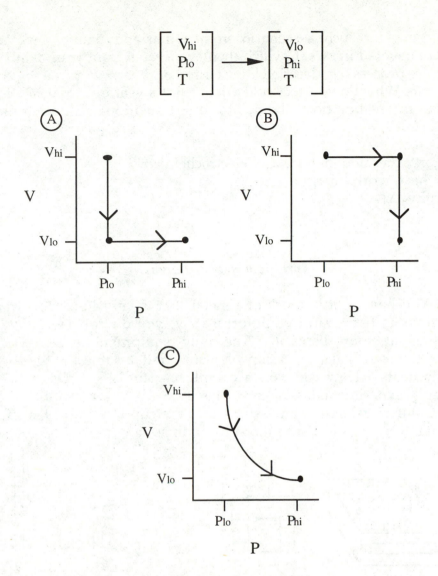

[category: thermodynamics and electrochemistry]
[topic: heat, work, energy]
[difficulty: A]

Cold Sweats
(problem seventy-seven)

In a variety of outdoor sports such as hiking and skiing, you have to stay warm and dress appropriately. Just as it is important not to underdress, it is very important not to overdress, because if you wear overly heavy clothing you can sweat. And when the sweat evaporates from your skin, you can get really cold.

Consider such a scenario from a thermodynamic perspective: You are covered in sweat, which then evaporates from your skin. How can this process be described in terms of heat, work, and energy changes? Why do you feel cold when your sweat evaporates? Is this process of evaporation (liquid to gas transition) endothermic or exothermic?

[category: thermodynamics and electrochemistry]
[topic: heat, work, energy]
[difficulty: A]

Hot Hot Hot
(problem seventy-eight)

You have a little block of a metal that is very hot. You want to determine its temperature. Unfortunately, you do not have any fancy machines at your disposal. The only equipment you have is a thermometer, a balance, a cup of water, and a table of all relevant specific heats. How can you accomplish your task? Describe your answer experimentally, and justify your reasoning with thermodynamics! You can assume that the cup is a perfect insulator, and that no energy is lost to the surroundings.

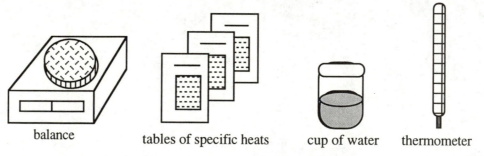

balance   tables of specific heats   cup of water   thermometer

temperature of block?

[category: thermodynamics and electrochemistry]
[topic: heat, work, energy]
[difficulty: B]

First Things First
(problem seventy-nine)

The First Law of Thermodynamics states that energy is conserved. In other words, the change in energy of the system is equal to the heat added to the system plus the work done on the system. What is the mathematical formulation of the First Law?

Imagine you have a vessel of volume V_1 containing an ideal gas at temperature T_1. You then place this vessel in a colder room of temperature T_2. The vessel's temperature starts to drop. What are the signs for the system's q, w, and E for this process? How can you express the magnitudes of each of these quantities?

You then compress the gas in a reversible manner using a piston until it reaches a smaller volume V_2. You make sure that during this process there is no temperature change (it is an isothermal process). What are the signs for the system's q, w, and E for this process? How can you express the magnitudes of each of these quantities?

[category: thermodynamics and electrochemistry]
[topic: heat, work, energy]
[difficulty: B]

What's Hot? What's Not?
(problem eighty)

What does it mean for a substance to feel "hot" or "cold"? You are interested in the subtleties of thermodynamics and decide to have a conversation with one of your colleagues. He snorts, "Why bother thinking about something so simple? Just measure the temperature with a thermometer. A high temperature means it will feel hot, and a low temperature means that it will feel cold. What a ridiculous question! Am I missing anything?"

You calmly present to him three blocks of substances with thermometers protruding from their interiors. The first block is wood, the second granite, and the third copper (see illustration). You cover the thermometers and ask him to close his eyes, touch the blocks with his hands, and comment on which one feels the warmest. He does so and says, "The first block is the warmest and the third one is the coolest. Of course this means that the first one has the highest temperature and the third one has the lowest temperature."

You show him the thermometers, and all read exactly 18°C. He is speechless. He feels the blocks again and cannot understand why the wood feels the warmest and the copper feels the coldest.

Explain to your colleague what is going on and why. Also, explain which would feel the warmest and coolest if they were all at 50°C.

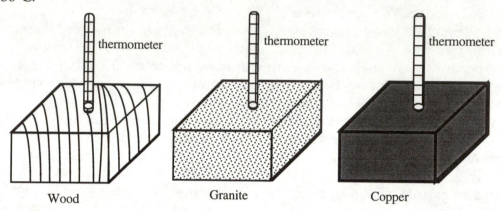

Wood   Granite   Copper

[category: thermodynamics and electrochemistry]
[topic: heat, work, energy]
[difficulty: B]

Cool It Down!
(problem eighty-one)

On a hot summer morning, you purchase an air conditioner and place it in the center of your room. Assume that your room is sealed and that a fan mixes the air in the room well. You let the air conditioner run all day long, and in the late afternoon you come back. What has happened to the temperature of the room and why? Justify with a qualitative and a quantitative argument.

[category: thermodynamics and electrochemistry]
[topic: heat, work, energy]
[difficulty: C]

Mountain Madness
(problem eighty-two)

People have long been attracted to mountain climbing for its beauty and challenge. A number of dangers, however, await these most adventurous of souls. For example, when an air mass passes over a mountain range, it can be compressed very rapidly when it reaches the far side. This is called a chinook and is dangerous for mountain climbers on the "far" side of the mountain. It occurs with very little transfer of heat and thus can be approximated as an adiabatic compression of an ideal gas. What happens to the temperature of the air on the "far" side of the mountain? Why? Why is this dangerous?

[category: thermodynamics and electrochemistry]
[topic: heat, work, energy]
[difficulty: C]

Diamonds Are Forever
(problem eighty-three)

Are diamonds really forever? At least James Bond thinks so. You look up in a textbook that when carbon and oxygen combine to form carbon dioxide, the reaction releases more free energy when diamond is used than when graphite is used as the carbon source. Based on this information, determine whether or not diamond will spontaneously decompose into graphite. If it does, then why do we not see our diamond rings becoming pencil points? If it does not, then does this imply that diamonds will last forever?

[category: thermodynamics and electrochemistry]
[topic: enthalpy, entropy, free energy]
[difficulty: A]

Doing the Impossible
(problem eighty-four)

In order to build large macromolecules and grow, a cell must perform many chemical reactions that involve positive changes in Gibbs free energy. That is, they are inherently *nonspontaneous*. How can these reactions occur? One strategy the cell uses is to have a large number of available molecules of adenosine triphosphate (ATP). ATP can be hydrolyzed to form adenosine diphosphate (ADP) plus inorganic phosphate ("P_i", or HPO_4^{2-}). This reaction is spontaneous: ATP —> ADP + P_i.

Explain in thermodynamic terms how a cell could utilize this to make a nonspontaneous reaction spontaneous. Include Hess' law in your explanation.

[category: thermodynamics and electrochemistry]
[topic: enthalpy, entropy, free energy]
[difficulty: A]

Everything in Disarray
(problem eighty-five)

Two common notions of entropy define it as a measure of "disorder" of the system or as a measure of the number of "microstates" that can describe the system. In each of the following examples, determine which system has a higher entropy. Justify each with an argument based of the "disorder" definition as well as the "microstates" definition:

(i) a royal flush or four aces, as a five-card poker hand;

(ii) a mole of liquid water or a mole of water vapor (at the same temperature);

(iii) one mole of N_2 mixed uniformly with one mole of O_2, or two moles of NO;

(iv) a gas system involving helium and hydrogen separated by a barrier, or the same system after the barrier is removed;

(v) a small sprout or a large oak tree. Consider this last one in terms of the Second Law of Thermodynamics. Does a tree sapling growing into a big tree violate this law? Explain.

[category: thermodynamics and electrochemistry]
[topic: enthalpy, entropy, free energy]
[difficulty: A]

Molecular Machines
(problem eighty-six)

Many of the molecular "machines" that function in living cells are part of a class of macromolecules called proteins. Proteins consist of a linear polymer of hundreds of amino acids, each of which can be either hydrophobic or hydrophilic in character. Expressed inside a cell, this linear string of amino acids spontaneously folds into a three-dimensional conformation that could be thought of as a compact ball of yarn (see illustration).

Exactly how the linear string of amino acids folds into its precise and defined three-dimensional complex structure represents one of the major unsolved problems facing biochemists today. This is known as the "protein folding problem". It is thought that the amino acids fold in a way such that the overall energy is minimized. Assume for the moment that the amino acids can be either hydrophobic or hydrophilic. (This is a simplification, since as you will learn if you study biochemistry, that there are 20 amino acids with a whole array of properties.) Would you expect the external surface of the folded protein to contain predominantly hydrophobic or hydrophilic amino acids? How about in the interior of the protein?

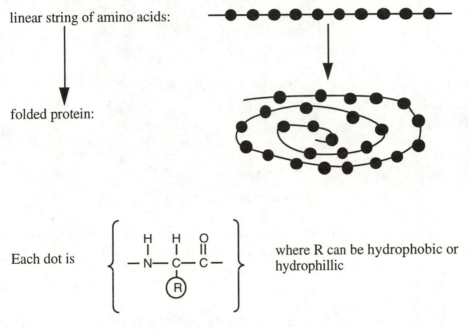

linear string of amino acids:

folded protein:

Each dot is {   } where R can be hydrophobic or hydrophillic

[category: thermodynamics and electrochemistry]
[topic: enthalpy, entropy, free energy]
[difficulty: A]

Cages of Water
(problem eighty-seven)

Review the protein folding discussion in the previous question. Think of the folding of a linear string of amino acids into a three-dimensional protein as a chemical reaction. Such a reaction can be written: Unfolded Linear String —> Folded Globular Protein.

Now think of the thermodynamics of the reaction. You know that the folding happens spontaneously and involves the formation of many stabilizing hydrogen bonds and salt bridges. Based on the above discussion is the change in Gibbs free energy (ΔG) positive or negative for the folding reaction? How about the enthalpy (ΔH)?

The entropic changes involved with the folding reaction involve two effects. First, the unfolded string of amino acids can be thought of as "disordered" and the folded protein with its defined and rigid three-dimensional structure as much more "ordered". However, a second effect predominates. Water molecules form very large, ordered "cages" around hydrophobic surfaces. These cages involve much more regular and rigid hydrogen bonding patterns than exist in pure water, and these cages are lost upon protein folding. Why does folding make the cages of water molecules disappear? What does this imply for the overall entropic (ΔS) change for the folding reaction? Is ΔH or ΔS (or both) the driving force for protein folding?

[category: thermodynamics and electrochemistry]
[topic: enthalpy, entropy, free energy]
[difficulty: B]

Cracking the Egg
(problem eighty-eight)

Humpty-Dumpty sat on a wall. Humpty-Dumpty had a great fall. All the king's horses and all the king's men couldn't put Humpty together again.

Consider this nursery rhyme in terms of the Second Law of thermodynamics. Is it true that Humpty absolutely cannot be put together again? In other words, is it forbidden by the laws of thermodynamics? If so, justify. If not, state what would be necessary to put Humpty together again.

[category: thermodynamics and electrochemistry]
[topic: enthalpy, entropy, free energy]
[difficulty: B]

Chemical Glue!
(problem eighty-nine)

You place solid barium hydroxide octahydrate [$Ba(OH)_2 \cdot 8H_2O$] with solid ammonium nitrate [NH_4NO_3] in a flask. After a minute of gentle shaking, a gas is evolved from the flask, and some liquid forms.

You place the flask on a wet piece of paper, and the paper sticks to the bottom of the flask!

What was the gas and liquid? Why did the paper stick to the flask? (Think of a simple explanation for this...) What is the entropic change in this reaction?

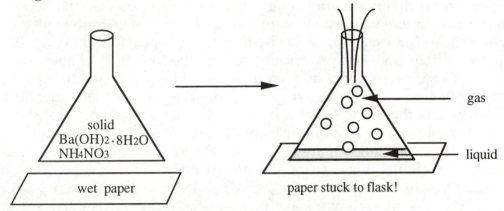

solid
$Ba(OH)_2 \cdot 8H_2O$
NH_4NO_3

wet paper   paper stuck to flask!

gas

liquid

[category: thermodynamics and electrochemistry]
[topic: enthalpy, entropy, free energy]
[difficulty: B]

Sherlock
(problem ninety)

Chemistry is often like detective work. Given only a few clues, you must combine them in the right way in order to figure out what you are looking for. Simple formula manipulations often can become powerful tools when combined appropriately.

Consider liquid diatomic bromine. As you know, in any liquid some molecules escape into the gaseous phase at the liquid-gas interface, creating a vapor pressure. You want to make a relatively good estimate of the vapor pressure of liquid bromine at room temperature, but you are only given three pieces of information: (i) the standard enthalpy of formation of gaseous diatomic bromine, (ii) the normal boiling point of bromine, and (iii) the fact that enthalpy and entropy changes do not strongly depend on temperature. Is this feat possible? If so, demonstrate this. If not, explain which piece of information is missing. (Do not use Trouton's Rule.)

[category: thermodynamics and electrochemistry]
[topic: enthalpy, entropy, free energy]
[difficulty: C]

Yin-Yang
(problem ninety-one)

The ancient Chinese Yin-Yang philosophy states that there is a fundamental dualistic nature to the forces of the universe. These two forces are complementary yet opposing and define the complexities of the world. For example, the male-female, lord-peasant, parent-child, and day-night divisions are formulations of the Yin-Yang philosophy.

A confused student decided that this philosophy directly opposes the Second Law of thermodynamics. Yin-Yang states that order and separation is the natural tendency of systems; thermodynamics states that chaos and disorder are the fate of the universe. In order to defend his position, the confused student set up the following experiment. An equimolar mixture of gas A and gas B is placed in one chamber, which is connected to another chamber by a pinhole. The two gases are both ideal, but B's molar mass is much greater than A's molar mass. Approximately half the gas is allowed to effuse spontaneously. Because light particles effuse faster than heavier ones, the gas in the original chamber is enriched in B, and the gas in the second chamber is enriched in A (see illustration on next page).

The student proudly claims before his audience: "There you have it, ladies and gents, the Yin-Yang philosophy hereby disproves the science of thermodynamics. A spontaneous process leads to the partial separation of two gaseous components, thus spontaneously decreasing entropy! Forbidden by the Second Law, it is what must be expected by the Yin-Yang ordering of the universe into a dualism of large and small objects."

Please critique this experiment and the conclusion drawn.

Two Contrasting Philosophies:

Yin-Yang   vs.   **Thermodynamics**

order   disorder
separation   chaos

gas A: low molecular weight

gas B: high molecular weight

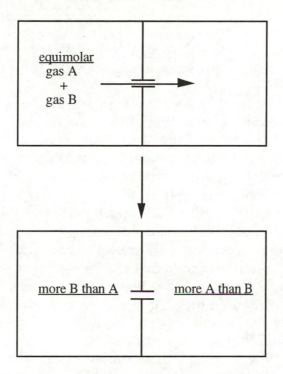

[category: thermodynamics and electrochemistry]
[topic: enthalpy, entropy, free energy]
[difficulty: C]

Turn Up the Heat
(problem ninety-two)

Hess's Law states that when chemical reactions can be added together, their reaction enthalpies are simply additive. This can be an extremely powerful tool in creating theoretical reaction cycles. Assume that you know the enthalpy of sublimation for graphite and the bond dissociation enthalpy for molecular oxygen. Your goal is to determine the enthalpy of atomization of carbon dioxide (the heat required to break one mole of gaseous carbon dioxide into one mole of gaseous carbon atoms and two moles of gaseous oxygen atoms). You have a supply of graphite, a quick mind, and a constant volume bomb calorimeter. Outline a procedure to solve the problem, and justify mathematically how you would compute the answer.

[category: thermodynamics and electrochemistry]
[topic: enthalpy, entropy, free energy]
[difficulty: C]

Really, Really Cold
(problem ninety-three)

The usual concept of absolute temperature is that it can only be positive. Temperature can be thought of as a measure of the average kinetic energy of molecules, and thus faster moving molecules have higher temperature. How fast would molecules at absolute zero be moving? Under this definition of temperature, why can you never have a negative absolute temperature?

An alternative definition (and perhaps the fundamental definition in some instances) of absolute temperature is that it is equal to the rate of change of energy with respect to entropy: T=dE/dS. From this follows the common result that the energy change in a constant-volume system is equal to the temperature times the change in entropy. Using this definition, if we only consider thermal energy, then we can derive a relation between the temperature and the common notion of "hotness or coldness".

Consider now a system that consists only of a lattice of small magnets in which kinetic energy is negligible. If we impart magnetic energy to the system by turning on a strong magnetic field, then the magnets line up their polarities (see illustration). Does the higher energetic state have a higher or lower entropy? What does this imply about the absolute temperature of the higher energetic state?

$$T = \frac{dE}{dS}$$

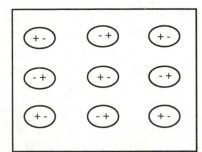

no magnetic field
random alignment of magnets

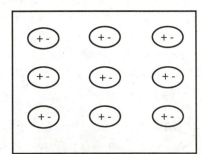

strong magnetic field/energy
all magnets aligned

[category: thermodynamics and electrochemistry]
[topic: enthalpy, entropy, free energy]
[difficulty: C]

Speeding Bullets
(problem ninety-four)

Stibnite is an ore of antimony used in manufacturing bullets. It is usually found in the form of a trisulfide and can be purified by a reaction with free iron to produce free antimony and ferrous sulfide. From your knowledge of oxidation states and electrochemistry, write the chemical equation for this process and balance it. Then identify which compound is reduced, which is oxidized, which is the reducing agent, and which is the oxidizing agent. Also write the two half reactions and the oxidation states of each atom in your reaction. Note that Sb forms a +3 ion and the oxidation number of S is -2 both in the reactants and in the products.

[category: thermodynamics and electrochemistry]
[topic: electrochemistry]
[difficulty: A]

Iron Breathing
(problem ninety-five)

In order to sustain life in higher organisms, oxygen must be transported via the bloodstream to all cells. Oxygen is transported in the body by red blood cells, which have many large protein molecules known as hemoglobin. An iron atom is wedged in four key locations of each hemoglobin and is directly responsible for binding oxygen and carrying it to various tissues. The iron atom can be in the ferrous or ferric oxidation state, and the resulting hemoglobins are called ferrohemoglobin and ferrihemoglobin. Ferrohemoglobin can bind oxygen, but ferrihemoglobin cannot. Often ferrihemoglobin results from diseases or various pathologies, and causes insufficient oxygen delivery to cells (causing a person to have a blue-ish appearance known in medical terminology as "cyanosis").

Which oxidation states of Fe are responsible for ferrohemoglobin and ferrihemoglobin? Which one is the oxidized form and which one is the reduced form? If a cellular molecule had the function of converting ferrihemoglobin to ferrohemoglobin, would this converting molecule be termed an oxidizing or reducing agent?

Ferrohemoglobin
iron (ferrous)
can bind oxygen
associated with health

Ferrihemoglobin
iron (ferric)
cannot bind oxygen
associated with disease

[category: thermodynamics and electrochemistry]
[topic: electrochemistry]
[difficulty: A]

Priceless Photos
(problem ninety-six)

Chemical reactions lie at the heart of the field of photography. A simple method of photography basically forms a "frozen image" of an object. Given the information below, your task is to develop a protocol for producing an image of an object.

When exposed to light, the ferrioxalate ion, $Fe(C_2O_4)_3^{3-}$, undergoes a redox reaction: the iron is reduced, and some (one-sixth) of the oxalate ions are oxidized to form carbon dioxide. Fe in a +2 oxidation state (and none other) can react with potassium ferricyanide, $K_3Fe(CN)_6$, to form free potassium ions and a ferroferricyanide complex, $KFeFe(CN)_6$, that is deep blue in color. All other compounds are colorless. You have a few petri dishes, a few pieces of filter paper, a strong lamp, and bottles of iron (III) nitrate, oxalic acid ($H_2C_2O_4$), potassium ferricyanide, and water. Note that iron (III) nitrate and oxalic acid form ferrioxalate ions in solution. Write out balanced chemical reactions for the formation of ferrioxalate ions, the redox reaction described above, and the formation of the blue complex. Using these reactions, describe how you could use them to make a negative print of an object. This experiment is simple enough for any first-year chemistry student.

petri dishes
filter paper
strong lamp
iron (III) nitrate   —>   photography
oxalic acid
potassium ferricyanide
water

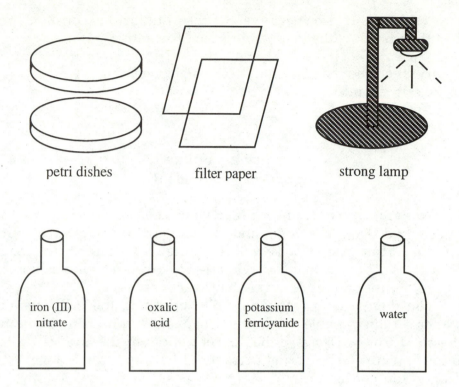

petri dishes   filter paper   strong lamp

iron (III) nitrate   oxalic acid   potassium ferricyanide   water

[category: thermodynamics and electrochemistry]
[topic: electrochemistry]
[difficulty: B]

Silly Cells
(problem ninety-seven)

A galvanic (or voltaic) cell uses a spontaneous chemical reaction in order to produce an electric current. You decide to construct a zinc-aluminum galvanic cell. One electrode consists of a bar of zinc in a molar solution of zinc nitrate, and the other electrode consists of a bar of aluminum in a molar solution of aluminum nitrate. You connect the electrodes with a wire and the solutions with a salt bridge. You are told by a friend that zinc has a higher standard reduction potential than aluminum. Sketch this cell, labeling the electrodes as "anode" and "cathode". Which is considered "+" and which is "-"? What is the direction of electron flow? What chemical reactions are occurring at each electrode? What is the overall redox reaction? After a period of time, will the bar of aluminum and the bar of zinc become heavier, lighter, or experience no change in mass? Is the free energy change for this reaction positive or negative? Is the equilibrium constant greater or less than one? Is the voltage positive or negative?

[category: thermodynamics and electrochemistry]
[topic: electrochemistry]
[difficulty: B]

Varying Volts
(problem ninety-eight)

You are experimenting with galvanic cells and decide to set up a standard hydrogen electrode and a standard iron/iron (II) nitrate electrode. Your friend tells you that the Fe^{2+} ion has a negative standard reduction potential. What is the overall cell reaction, and what are the individual electrode half-reactions?

You then want to see what will happen to the cell potential if you alter certain variables. What will happen if you decrease the mass of the solid iron? Decrease the concentration of the iron (II) ions in solution? Increase the partial pressure of hydrogen gas? Increase the hydrogen ion concentration?

[category: thermodynamics and electrochemistry]
[topic: electrochemistry]
[difficulty: B]

Flowing Ions
(problem ninety-nine)

You have a tray of water, which you separate into halves by a semipermeable barrier. On the left side you add a high concentration (C_1) of KCl, and on the right side you add a little bit (C_2) of KCl. The barrier will allow potassium ions through but not anything else (not the chloride ions). Initially potassium will flow from the left to right side due to the free energy of the chemical gradient. As this happens, however, an electrical counteractive force will develop, since the movement of potassium will generate a net negative charge on the left and a net positive charge on the right. Eventually, an equilibrium will be reached when there is no net flux of potassium across the barrier (see illustration). What is an expression for the free energy of a chemical concentration difference across the barrier? What is an expression for an electrical charge difference across the barrier? What is the relative magnitude of these opposing forces initally? At equilibrium? What is an expression for the voltage generated across the barrier at equilibrium?

[For those with an interest in cell biology: This problem is analogous to the setup of a living cell's membrane. Inside the cell there is a high concentration of potassium, and outside there is a low concentration of potassium. The cell membrane is largely impermeable to most ions except for potassium, for which there are specific channels. The concentration difference leads to a voltage across the cell membrane known as the "membrane potential".]

Initially:

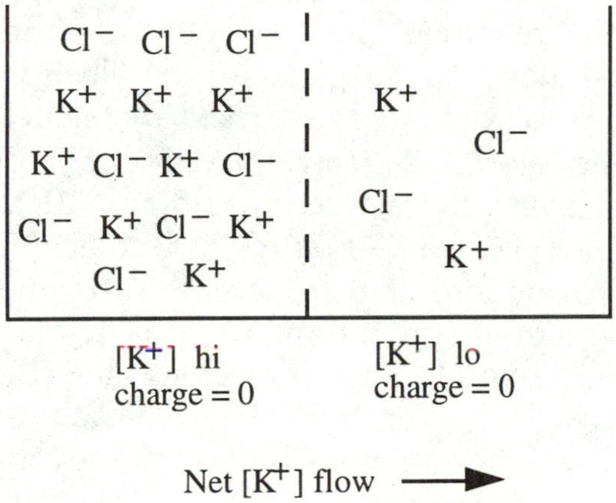

Cl^- Cl^- Cl^- | K^+
K^+ K^+ K^+ | K^+ Cl^-
K^+ Cl^- K^+ Cl^- | Cl^-
Cl^- K^+ Cl^- K^+ | K^+
Cl^- K^+ |

$[K^+]$ hi $[K^+]$ lo
charge = 0   charge = 0

Net $[K^+]$ flow ⟶

Equilibrium:

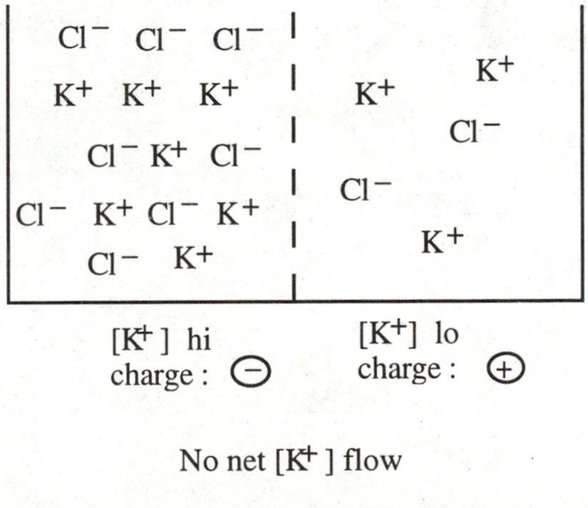

Cl^- Cl^- Cl^- | K^+
K^+ K^+ K^+ | K^+
Cl^- K^+ Cl^- | Cl^-
Cl^- K^+ Cl^- K^+ | Cl^-
Cl^- K^+ | K^+

$[K^+]$ hi $[K^+]$ lo
charge : ⊖   charge : ⊕

No net $[K^+]$ flow

[category: thermodynamics and electrochemistry]
[topic: electrochemistry]
[difficulty: C]

Waterfall of Energy
(problem one hundred)

One application of redox chemistry is a series of reactions that occurs in organelles of cells called mitochondria. The breakdown of food and nutrients yields molecules of NADH. The NADH molecules enter a chain of reactions in which electrons are transferred from NADH ultimately to molecular oxygen. The two half-reactions for the overall stoichiometric reaction are:

oxidation: $NADH \longrightarrow NAD^+ + H^+ + 2e^-$
reduction: $(1/2) O_2 + 2H^+ + 2e^- \longrightarrow H_2O$

Oxygen does not, however, immediately pick up the electrons released by NADH. A whole chain of molecules and redox reactions are involved. Each one in turn picks up the electrons (reduction) and then delivers them (oxidation) to the next molecule in the chain (see illustration). The electrons from NADH are first transferred to complex I, then to coenzyme Q, then to complex III, then to cytochrome c, then to complex IV, and finally to molecular oxygen. Each individual redox reaction step proceeds spontaneously, and each releases a small amount of energy (that is captured in a complex process not described here to form molecules of ATP).

Write down the overall stoichiometric reaction. In each individual redox reaction step and for the overall stoichiometric reaction, is the voltage positive or negative? Is the Gibbs free energy change positive or negative? Order the following in terms of increasing reduction potential: complexes I, III, IV, NAD, cytochrome c, coenzyme Q, and molecular oxygen. Which ones are the best electron donors and the best electron acceptors?

The Electron Transport Chain:

NADH —> Complex I —> Coenzyme Q —> Complex III —>

Cytochrome c —> Complex IV —> Molecular Oxygen

Electron Transfers:

Net Reaction:

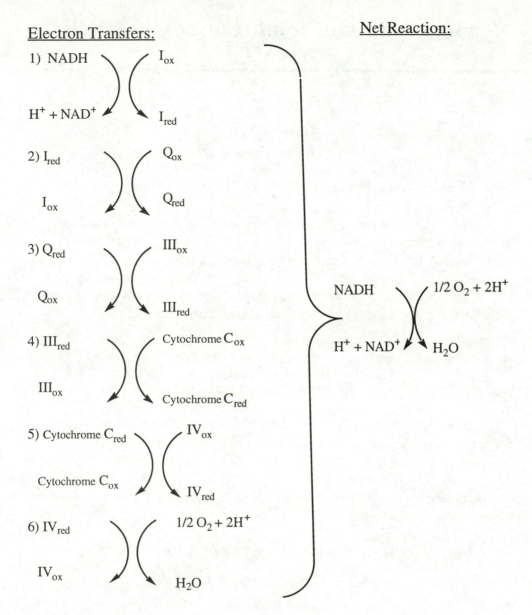

1) NADH

I_{ox}

$H^+ + NAD^+$

I_{red}

2) I_{red}

Q_{ox}

I_{ox}

Q_{red}

3) Q_{red}

III_{ox}

Q_{ox}

III_{red}

4) III_{red}

Cytochrome C_{ox}

III_{ox}

Cytochrome C_{red}

5) Cytochrome C_{red}

IV_{ox}

Cytochrome C_{ox}

IV_{red}

6) IV_{red}

$1/2\ O_2 + 2H^+$

IV_{ox}

H_2O

NADH

$1/2\ O_2 + 2H^+$

$H^+ + NAD^+$

H_2O

[category: thermodynamics and electrochemistry]
[topic: electrochemistry]
[difficulty: C]

Thermodynamics and Electrochemistry — HINTS

Stately Paths
(problem seventy-six)

Hint 1: What is the First Law of Thermodynamics?

Hint 2: What does the First Law imply regarding the conversions between heat, work, and energy? Can energy be created or destroyed? Can forms of energy be interconverted?

Hint 3: Which quantities are only dependent on the initial and final states of the system, "state" or "path" functions?

Cold Sweats
(problem seventy-seven)

Hint 1: What is the First Law of Thermodynamics? How is the fact that energy is conserved related to this problem?

Hint 2: For evaporation to occur, does the liquid absorb or release energy? What are the q, w, and E changes in the evaporation process?

Hint 3: What heat transfer is going on here?

Hot Hot Hot
(problem seventy-eight)

Hint 1: What is the relationship between energy changes and specific heats?

Hint 2: How can the law of conservation of energy be applied here? What energy equality could be relevant?

Hint 3: Why do you need the scale?

First Things First
(problem seventy-nine)

Hint 1: What are the thermodynamic definitions of heat, work, and energy?

Hint 2: What are the conventions for the signs of q and w? What does a positive/negative value mean? (Note that different texts have different conventions on this.)

Hint 3: What is the relationship between heat and the heat capacity of a gas? What is the relationship between work P-V changes?

What's Hot? What's Not?
(problem eighty)

Hint 1: What is the fundamental definition of heat? Temperature? How are they related?

Hint 2: When you feel that an object is hot or cold, what process is taking place? How can you describe this process in thermodynamic terms?

Hint 3: Are there other properties of a substance that affect how hot or cold it feels besides temperature?

Cool It Down!
(problem eighty-one)

Hint 1: What does an air conditioner do? How does it work? Use thermodynamic terminology.

Hint 2: What is the maximum efficiency possible for a machine?

Hint 3: What does the law of conservation of energy say about the available useful work output of a machine?

Mountain Madness
(problem eighty-two)

Hint 1: What is the thermodynamic definition of "adiabatic"? How does this simplify the First Law of Thermodynamics?

Hint 2: When the gas is compressed, is work being done *on* the system or *by* the system?

Hint 3: How is the (kinetic) energy of a gas related to temperature?

Diamonds Are Forever
(problem eighty-three)

Hint 1: What are the reactions for the formation of carbon dioxide from diamond and graphite? Can they be combined?

Hint 2: What are the conditions for spontaneity?

Hint 3: What other factors, besides the criteria for spontaneity, are important in determining whether a reaction will proceed or not?

Doing the Impossible
(problem eighty-four)

Hint 1: What is Hess' law?

Hint 2: If two reactions are added together, what is the overall free energy change?

Hint 3: How is free energy related to sponteneity?

Everything in Disarray
(problem eighty-five)

Hint 1: Which system is more probable to exist, one with high entropy (large number of microstates) or one with low entropy (small number of microstates)?

Hint 2: Which system is more ordered in terms of inter-molecular structures, a liquid or a gas? How can parts (iii) and (iv) be thought of as analogous?

Hint 3: What does the Second Law say about processes that decrease the entropy of the system?

Molecular Machines
(problem eighty-six)

Hint 1: Imagine the protein as a "ball of yarn." What would amino acids on the exterior of the protein interact with?

Hint 2: What would amino acids in the interior of the protein interact with?

Hint 3: How do hydrophobic and hydrophilic substances interact to minimize energy? (Think of oil droplets in water.)

Cages of Water
(problem eighty-seven)

Hint 1: What is the relationship between ΔG and sponteneity of a reaction?

Hint 2: What is the relationship between ΔH and hydrogen bond or salt bridge formation?

Hint 3: What is the relationship between ΔS and the cages of water molecules?

Cracking the Egg
(problem eighty-eight)

Hint 1: What is the sign of the entropy change for Humpty falling and breaking into pieces? For Humpty being put back together again?

Hint 2: For these processes, define the system, the surroundings, and the universe?

Hint 3: What's the Second Law of Thermodynamics? Be careful about whether it is referring to the system, surroundings, or universe in its conclusions about entropy changes of processes.

Chemical Glue!
(problem eighty-nine)

Hint 1: What is the balanced reaction for barium hydroxide octahydrate and ammonium nitrate?

Hint 2: Why is it necessary for the paper to be wet for the "glue" to work?

Hint 3: What is the relationship between the free energy, enthalpy, and entropy changes in a chemical reaction (at constant temperature and pressure)? What are the signs of each for this reaction?

Sherlock
(problem ninety)

Hint 1: For a vaporization reaction, what is the expression for the equilibrium constant? How does this depend on free energy?

Hint 2: How does free energy depend on entropy, enthalpy, and temperature? Do you know the enthalpy change of the reaction?

Hint 3: Can you calculate the entropy change of the reaction? At the boiling point, what is the free energy change of the reaction?

Yin-Yang
(problem ninety-one)

Hint 1: What are the relative rates of effusion of the light and heavy gases?

Hint 2: What else is occurring to the gas besides separation?

Hint 3: What does the Second Law say in terms of entropy changes of the system, surrounding, and universe for a process?

Turn Up the Heat
(problem ninety-two)

Hint 1: Write the reactions associated with atomization, sublimation, and bond dissociation. What are the phases of each substance? How can these equations be related?

Hint 2: How can the various enthalpies of reaction be related? What is the reaction that would be necessary to complete the cycle? How can the enthalpy of this reaction be determined experimentally?

Hint 3: In a bomb calorimeter, what quantity are you measuring? Heat, energy, enthalpy, or something else?

Really, Really Cold
(problem ninety-three)

Hint 1: What is the relationship between the energy and temperature of an ideal gas? How does this place an absolute lower limit on the temperature?

Hint 2: Rewrite all the sentences in the second paragraph in mathematical formulas. What is the fundamental thermodynamic definition of temperature?

Hint 3: In a normal system of matter, would increasing the thermal energy increase or decrease the randomness of the system? In the magnetic lattice, would increasing the magnetic energy increase or decrease the randomness?

Speeding Bullets
(problem ninety-four)

Hint 1: How can the knowledge of the oxidation numbers be used to determine the empirical formulas of all the compounds?

Hint 2: What is the change in oxidation number of the antimony in the stibnite and the free antimony? What is the change in oxidation number of the iron?

Hint 3: What are the definitions of oxidation and reduction in terms of changes in number of electrons and oxidation states?

Iron Breathing
(problem ninety-five)

Hint 1: What are the two common oxidation states of Fe?

Hint 2: What is the definition of oxidized and reduced in terms of oxidation states?

Hint 3: Is an oxidizing agent itself oxidized or reduced when it functions?

Priceless Photos
(problem ninety-six)

Hint 1: What are the only possible oxidation states of iron? This should remove any ambiguity regarding the reactions.

Hint 2: There is a key difference between the reaction that is sensitive to light and the development reaction. Which is which?

Hint 3: Besides describing and ordering the chemical reactions, what additional factors would be necessary in achieving experimental success (handling, speed, timing, etc.)?

Silly Cells
(problem ninety-seven)

Hint 1: For the spontaneous reaction in the galvanic cell, which metal must be reduced and which must be oxidized?

Hint 2: What does the spontaneity of the reaction tell you about the free energy, equilibrium position, and voltage of the cell?

Hint 3: How are the relations between free energy, equilibrium, and voltage represented mathematically?

Varying Volts
(problem ninety-eight)

Hint 1: What is the standard hydrogen electrode? What are standard conditions for other electrodes?

Hint 2: Would a spontaneous reaction have a positive or negative overall cell potential? Is the Fe^{2+} ion reduced or the Fe metal oxidized?

Hint 3: How does the Nernst equation relate the cell potential to nonstandard conditions?

Flowing Ions
(problem ninety-nine)

Hint 1: How is free energy related to the equilibrium constant? How would you express the equilibrium constant for the reaction of a potassium ion moving from the left to the right side?

Hint 2: How is free energy related to the voltage across the barrier? The voltage is the potential difference for the reaction of a potassium ion moving from the left to the right side.

Hint 3: What is Nernst equation? How is it related to the equilibrium situation of this problem?

Waterfall of Energy
(problem one hundred)

Hint 1: What are the conditions for sponteneity in terms of Gibbs free energy?

Hint 2: What is the relationship between Gibbs free energy and electric potential?

Hint 3: In a redox reaction, which species is reduced, the one with the higher or lower reduction potential?

Chapter 7 — Quantum Theory, Atomic Structure, and Periodic Trends

Fun in the Sun
(problem one hundred one)

Getting sunburned is dangerous as well as painful. Too much exposure to high energy electromagnetic radiation can lead to cellular DNA damage causing an increased risk for cancer. You know that DNA damage occurs if the energy of the incident electromagnetic radiation is above a certain threshold. What would you predict is worse based on your knowlegde of quantum theory, a weak beam of ultraviolet light or an intense beam of infrared light? Explain.

[category: quantum theory, atomic structure, and periodic trends]
[topic: quantum theory]
[difficulty: A]

Jumpy Electrons
(problem one hundred two)

A number of observable phenomena result from the transitions of electrons between energy states. These phenomena contributed greatly to the development of the modern conception of atomic structure.

Consider first the electronic transitions in the hydrogen atom. You have a hydrogen lamp that looks purple in color. Looking at this through either a prism or a diffraction grating separates this light into a series of lines known as the first four lines of the Balmer series (red, green, and two different blue lines). Why do you see only these colors? Compare this discrete spectra with the continuous spectra obtained by passing white light through a prism or grating. The Balmer series is only one of a number of such series of possible emissions from hydrogen. Why do you not see the others?

In addition to emission spectra, other observable properties due to "jumpy electrons" are fluorescence and phosphorescence. These are two closely related phenomena. In fluorescence, electrons in a substance move to an excited state upon absorption of energy. These electrons very rapidly fall back to the ground state, emitting photons of light and causing the material to "fluoresce". An example of this is a white cotton T-shirt "glowing" under ultraviolet light. When the excitation energy source is turned off, the material stops fluorescing. Why does this happen? In phosphorescence, a material continues to emit photons of light even after the excitation energy source is turned off. What does this imply about the rate of electronic transitions in phosphorescence compared with fluorescence?

[If you study physical chemistry or quantum mechanics, you will learn the fundamental differences between fluorescence and phosphorescence. Quantum theory only allows certain "transitions" between energy levels. The description of which transitions are allowed or forbidden are known as the selection rules. Fluorescence results from electronic transitions that are "allowed", whereas phosphorescence results from transitions that are "forbidden".]

[category: quantum theory, atomic structure, and periodic trends]
[topic: quantum theory]
[difficulty: A]

Schizophrenia
(problem one hundred three)

Every moving particle acts as a wave, and every electromagnetic wave has certain mass characteristics. The DeBroglie wavelength is a representation of the wave character of any massive object. Planck's energy quantization represents the energy associated with one photon of radiation, and according to Einstein, this energy is associated with a certain "effective mass". State mathematically and intuitively why a flashlight does not seem to emit "mass" particles and why a baseball does not appear like a "wave". Then describe at least one observation of electrons exhibiting "wave" characteristics and electromagnetic radiation exhibiting "mass" characteristics. What's different from the first set of examples?

The three formulas described above are often thought to form the core of an area in quantum theory called the wave-particle duality, which states that we cannot create a division between these two concepts. To show the intimate connection between these three formulas and their aesthetic completeness, pick any two formulas and derive the third one for a photon of light.

[category: quantum theory, atomic structure, and periodic trends]
[topic: quantum theory]
[difficulty: B]

Peeping Tom
(problem one hundred four)

According to the wave-particle duality, electrons must simultaneously be thought of as particles and waves. You can have a beam of electrons that act as a stream of particles, but there are certain experiments in which electrons exhibit the characteristics of a wave. For example, in the "double slit" experiment, a beam of electrons is focused on a plate that has two slits. Beyond the plate is a detector screen that records the numbers of electrons that hit at each point.

The double slit experiment cannot be described by the "particle" notion of an electron. Particles would be forced to go through either one slit or the other, creating two equal spots on the detector screen. A wave, on the other hand, would simultaneously go through both slits and create a complex interference pattern on the detector screen.

You turn on the detector screen and the electron beam and see an interference pattern. The electrons are behaving as waves!! Then you try to see how many electrons are going though each slit. You set up an additional detector at each slit and count that about half of the electrons are going through each slit. However, with the counters at each slit, you no longer see the same interference pattern on the detector screen. Now you see two main spots on the screen, suggesting that the electrons now act as particles!! Explain these observations by the wave and particle descriptions of electrons. Then comment on why the *act* of counting the electrons made them seem to switch from "wave" to "particle" nature.

[category: quantum theory, atomic structure, and periodic trends]
[topic: quantum theory]
[difficulty: B]

Curved Waves
(problem one hundred five)

You are studying the properties of waves. A wave front approaches a boundary that blocks the wave propagation except in the region of a single small opening (see illustration on next page). The waves propagate straight initially, but after the opening in the boundary, the waves propagate out in semicircles! Why does this happen? Explain in terms of Heisenebrg's Uncertainty Principle, applied to the position and momentum of the wave at the opening in the boundary.

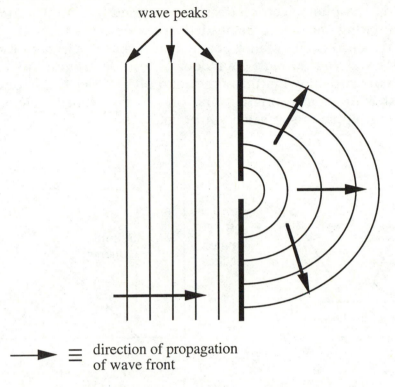

wave peaks

$\longrightarrow \equiv$ direction of propagation
of wave front

[category: quantum theory, atomic structure, and periodic trends]
[topic: quantum theory]
[difficulty: B]

To Catch a Thief
(problem one hundred six)

In 1905, Albert Einstein solved a famous chemistry and physics problem called the photoelectric effect. You believe that a thief wants to break into your apartment, and you want to catch him red-handed, using Al's brilliant discovery. Use the tools below and your knowledge of quantum theory to build a high-tech burglar alarm!

In the photoelectric effect, a metal can emit electrons when irradiated by electromagnetic radiation that has at least a certain frequency. This is called the threshold frequency. Light below this frequency has no effect on the metal; light with higher frequency, on the other hand, causes electrons to be ejected with kinetic energy equal to the energy of the incoming photon minus the energy required just barely to eject an electron (see illustration).

You have a block of metal, a high energy UV laser (capable of producing the photoelectric effect in your metal), a video camera, and as much wiring and stands as needed. Your object is to place the video camera (inconspicuously) in a position such that it will take the picture of anyone walking through your door. It can be programmed to be on when a current is flowing through it and off at other times, or off when a current is flowing through it and on at other times. How can you design your apparatus to catch your thief?

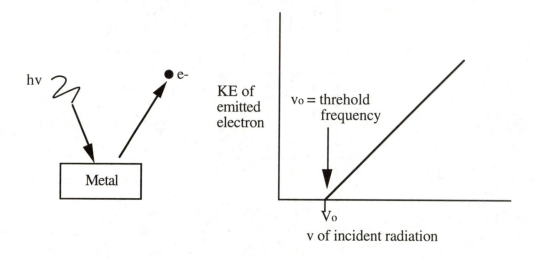

$$KE_{electron} = h\nu - h\nu_0$$

[category: quantum theory, atomic structure, and periodic trends]
[topic: quantum theory]
[difficulty: C]

Revolution!
(problem one hundred seven)

Most scientific explanations are based on existing scientific laws, which are unproven assertions that have not yet observed to be violated. Occasionally, a new breakthrough theory overthrows these laws and thus changes our entire method of thinking about the subject. Such a revolution occurred in the early 1900's in physics with the development of relativity and quantum mechanics. These new ideas challenged long accepted dogma and ushered the world into a new era of scientific understanding. In light of such revolutions, it is obvious that there is no such thing as absolutism in science, only self-consistency. We can only claim that theories are consistent with all phenomena that have been observed.

Consider the revolution from classical to quantum physics. In classical physics, energy levels were continuous, mass was conserved, speed was unlimited, physical objects were not waves, and the position and momentum were independent of each other. Explain how each of these concepts was reformulated with relativity and quantum theory. How do they all depend on very large constants (such as the speed of light c) or very small constants (such as Planck's constant h)? Why is this essential for people not detecting relativistic or quantum effects in everyday life? Give an example of when each of these relativistic or quantum effects becomes significant.

Optional: What would our world be like if the numerical value of c was much smaller and the value of h was much bigger?

[category: quantum theory, atomic structure, and periodic trends]
[topic: quantum theory]
[difficulty: C]

Aufbau and Hund
(problem one hundred eight)

The quantum theory allows for the description of any electron in an atom by four quantum numbers. These quantum numbers are:

n = principal quantum number = 1, 2, 3, ...
l = azimuthal quantum number = 0, 1, ..., n-1
m_l = magnetic quantum number = -l, ..., 0, ..., +l
m_s = spin quantum number = +1/2 or -1/2

What are the possible values of the four quantum numbers for the following? Assume all atoms are in the ground state.

(i) The electron of a hydrogen atom;
(ii) The most readily ionizable electron of cesium;
(iii) One of the outermost electrons of iodine;
(iv) One of the outermost electrons of uranium;
(v) One of the d-electrons of chromium, with the knowledge that at least one of the other d-electrons has a positive spin.

[category: quantum theory, atomic structure, and periodic trends]
[topic: atomic structure]
[difficulty: A]

Noble and Inert
(problem one hundred nine)

For a long time, the elements in group VIII of the periodic table (He, Ne, Ar, Kr, Xe) were thought to be totally inert. Thus they became known as the noble gases, because they were thought not to react with anything. What is the electronic basis of this theory?

It was then shown that Xe can react with F and O, and that Kr can react with F. Why can fluorine and to a lesser extent oxygen, but not other elements, combine with these noble gases? Why do they combine more readily with the noble gases lower in the periodic table?

[category: quantum theory, atomic structure, and periodic trends]
[topic: atomic structure]
[difficulty: A]

Heavy Metal
(problem one hundred ten)

The transition metals are in the "*d*-block" of the periodic table. Consider four metals that are thought to have great monetary value: copper, silver, gold, and platinum. These have great historical significance throughout Western and Eastern civilizations as well as mystique and value in the modern world. What are the electronic structures of these four metals? State the principal, azimuthal, and magnetic quantum numbers of the last electron in each. What are the electronic structures of the most common ion(s) that can be formed from each atom?

[category: quantum theory, atomic structure, and periodic trends]
[topic: atomic structure]
[difficulty: A]

Untrodden Paths
(problem one hundred eleven)

All the elements with atomic numbers above 100 are very unstable and have very short lifetimes. There is a suggestion, however, that if element number 118 could be made, then it would be relatively stable. What would be its electron configuration, and why might it be more stable than the other elements with atomic numbers above 100? Why would it be still less stable than other elements in its group?

[category: quantum theory, atomic structure, and periodic trends]
[topic: atomic structure]
[difficulty: B]

Natural Attraction
(problem one hundred twelve)

Mr. Chem is studying the structure of the helium atom. "Okay," he says, "I understand that it contains a nucleus with two protons and two neutrons, surrounded by two electrons. But I'm confused. My brother, Mr. Phys, tells me that like charges repel and unlike charges attract. Thus, two protons should repel each other, and a proton and electron will attract each other (see illustration). Why, then, don't the protons fly apart, and why don't the electrons collapse onto the protons?"

You scratch your head for a moment and then reply, "Mr. Chem, do not worry. I have the answer ..."

What's your response?

Helium (stable)

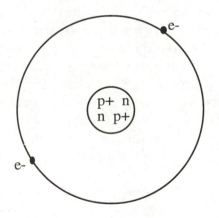

Two Hydrogen (H+) Ions

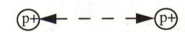

strong repulsion!

[category: quantum theory, atomic structure, and periodic trends]
[topic: atomic structure]
[difficulty: B]

Levels of Energy
(problem one hundred thirteen)

The Schrodinger equation for the hydrogen atom, a two body physics problem, can be solved exactly. Do the energy levels of orbitals in the hydrogen atom depend on the principal quantum number n? The angular quantum number l? The magnetic quantum number m?

With multi-electron atoms, the electron-electron repulsion creates mathematical difficulties that make only approximate numerical solutions possible. It also adds an additional complication to determining the energies of the orbitals. For multi-electron atoms, do the energy of orbitals depend on n, l, and m?

Why does Bohr's theory of atomic orbits correctly predict the energy levels of hydrogen (and single-electron ions) but no other atoms? Comparing the above two situations should make apparent the answer to this question.

[category: quantum theory, atomic structure, and periodic trends]
[topic: atomic structure]
[difficulty: C]

Orbitals and Orbits
(problem one hundred fourteen)

The classical conception of the atom is exemplified by the Bohr model of the atom. A central, dense nucleus is surrounded by orbiting electrons, analogous to the sun and planets in our solar system. The modern conception of the atom is the quantum mechanical version involving orbitals described by the wave function Ψ. What are the fundamental differences between the concept of classical electron orbits and quantum mechanical electron orbitals? How do the Bohr orbits conflict with the wave-particle duality and Heisenberg's uncertainty principle? How do the conception of orbitals take the wave nature of the electrons into account?

Atomic orbitals originate from the mathematical solutions of Schrodinger's wave equation, which is a description of the kinetic and potential energies of the electrons. The "Hamiltonian" function acting on the wave function Ψ equals a scalar multiple of Ψ representing energy (see equation on next page). Don't worry about the actual mathematics involved, as you will study this in the future if you take a course in quantum mechanics or physical chemistry.

You should understand that the solutions Ψ are standing wave functions that describe the electron's position and momentum as accurately as they are allowed by the principles of quantum mechanics. The square of the wave function, Ψ^2, is the *probability density* of finding an electron in a particular location. It follows, then, that the *probability* of finding an electron in a particular region of space dV is given by Ψ^2 dV, which can be integrated over any finite-sized region of three-dimensional space. The wave function solutions are mathematical functions (combinations of polynomial, trigonometric, and exponential functions) that are related to the regions of space in which it is likely to find an electron. They depend on three variables, n, l, and m, which entirely determine the physical characteristics of the standing wave. Thus the orbital's shape, size, and orientation can be entirely determined by these three variables.

For a 1s orbital, what are the values of n, l, and m? Sketch the probability density of this orbital in space, and show what shape is enclosed by a 90% probability confidence boundary. How many radial, angular, and total nodes does this orbital have? Answer the same questions for 2s, 2p, 3s, and 3p orbitals. Is a 3s electron necessarily further from the nucleus than a 2s or a 1s electron?

$$\mathbf{H}\,\Psi = E\,\Psi \qquad \text{Schrodinger Equation}$$
$$\Psi^2\,dV \qquad \text{Electron Probability Density}$$
$$\Psi = \Psi\,(n, l, m) \qquad \text{Wave Function Form}$$

[category: quantum theory, atomic structure, and periodic trends]
[topic: atomic structure]
[difficulty: C]

Are You Affable?
(problem one hundred fifteen)

Electron affinities follow more complex trends than ionization energy. Explain the following observations in terms of electronic structure: H, Li, Na, K, B, Al, and Ga have small, negative electron affinities; C, Si, Ge, N, P, and As have more negative electron affinities; O, S, and Se have very negative electron affinities; F, Cl, and Br have extremely negative electron affinities; and He, Ne, Ar, and Kr have positive electron affinities. What is the general trend of the elements with negative electron affinities? What property is shared between the elements with positive electron affinities that makes gaining an electron so unfavorable?

[category: quantum theory, atomic structure, and periodic trends]
[topic: periodic trends]
[difficulty: A]

Identity Crisis (I)
(problem one hundred sixteen)

I am one of the elements. I am a soft metal, and you can cut me with a knife, but watch out because I am highly reactive. I have the largest atomic radius of any element in my period. Within my group I have the largest atomic radius of the elements that have no d-electron shells filled. Who am I?

[category: quantum theory, atomic structure, and periodic trends]
[topic: periodic trends]
[difficulty: A]

Fashion Trends
(problem one hundred seventeen)

Consider the following ions/atoms: O^{2-}, F^-, Ne, Na^+, and Mg^{2+}. Order them in terms of increasing ionization energy. Also order them in terms of increasing radius.

Now consider the following atoms: O, F, Ne, Na, Mg. Order them in terms of increasing ionization energy. Also order them in terms of increasing radius.

Why is $IE_1 < IE_2 << IE_3$ for calcium whereas $IE_1 << IE_2 < IE_3$ for potassium? (IE_x stands for the x-th ionization energy.)

[category: quantum theory, atomic structure, and periodic trends]
[topic: periodic trends]
[difficulty: B]

Identity Crisis (II)
(problem one hundred eighteen)

I am one of the elements. I have a high electron affinity (highly negative value), and my atomic number is N. The element with atomic number N-1 has a lower ionization energy and a lower electron affinity (less negative value). The element with atomic number N+1 has a higher ionization energy and basically no electron affinity (positive value). I am toxic in my elemental state, but I am very commonly found in my nontoxic ionic state. Within my group, I have the second highest ionization energy. Who am I?

[category: quantum theory, atomic structure, and periodic trends]
[topic: periodic trends]
[difficulty: B]

Trinity
(problem one hundred nineteen)

Many atomic properties and trends can be studied by merely considering the first three elements in the periodic table. What are the electronic structures of H, He^+, He, Li^{2+}, Li^+, and Li? How many protons, neutrons, and electrons does each species have? Order them in terms of increasing ionization energy and in terms of increasing radius. Is the 1s shell a different size in H and Li?

[category: quantum theory, atomic structure, and periodic trends]
[topic: periodic trends]
[difficulty: C]

Identity Crisis (III)
(problem one hundred twenty)

I am one of the elements. I am a metal that can form cations. My atomic radius is smaller than the atomic radius of the element with one fewer protons, but my +1 ionic radius is larger than the +1 ionic radius of the element with one fewer protons. Of the two elements in my group that have great biological importance, I am the one with higher electronegativity. Who am I?

[category: quantum theory, atomic structure, and periodic trends]
[topic: periodic trends]
[difficulty: C]

Quantum Theory, Atomic Structure, and Periodic Trends — HINTS

Fun in the Sun
(problem one hundred one)

Hint 1: Where do UV and IR radiation fall in the electromagentic spectrum?

Hint 2: What is Planck's energy quantization equation?

Hint 3: What determines energy, intensity or frequency of the wave?

Jumpy Electrons
(problem one hundred two)

Hint 1: What is white light? What are its frequencies and wavelengths?

Hint 2: The first four lines of the Balmer series are visible, but the other lines and the other series are not. Are these transitions not occurring? Or are they occurring but undetectable by the naked eye?

Hint 3: How would the type of transition ("forbidden" or "allowed") determine the rate of the electron "jumping"?

Schizophrenia
(problem one hundred three)

Hint 1: What is the mathematical formulation of Planck's energy quantization, DeBroglie's wavelength, and Einstein's mass-energy formulas?

Hint 2: What is the value of Planck's constant? Very generally, is this value "large" or "small" in comparison with standard measurements?

Hint 3: What is the energy of a photon of light? What is its effective mass? What would be the DeBroglie wavelength of a "particle" of this mass?

Peeping Tom
(problem one hundred four)

Hint 1: Sketch the experiment outlined above. How would the electrons behave through the slits as particles? As waves?

Hint 2: Can the observer be separated from the experiment? Why does knowing the positions of the electrons at the slits matter? Why does this change the pattern seen on the detection screen?

Hint 3: Does a beam of electrons switch from being a wave to being a stream of particles, or does it represent some sort of hybrid of these two ideas?

Curved Waves
(problem one hundred five)

Hint 1: What is Heisenberg's Uncertainty Principle?

Hint 2: How certain is the position at the opening? How certain is the momentum at the opening?

Hint 3: The momentum is a vector quantity. What would certainty/uncertainty of momentum tell you about the direction of wave propagation?

To Catch a Thief
(problem one hundred six)

Hint 1: What is the mathematical representation of the photoelectric effect? What are two reasons why a ultraviolet laser would be preferable to a visible light laser for this situation?

Hint 2: What constitutes an electric current? How can the photoelectrode be used as part of an electric circuit?

Hint 3: How could you couple the physical act of walking through the doorway with an electrical response involving the photoelectric effect?

Revolution!
(problem one hundred seven)

Hint 1: What are the mathematical and conceptual formulations of energy quantization, mass-energy conversion, speed limits, matter waves, and the uncertainty principle?

Hint 2: Who were the pioneer scientists who formulated these concepts?

Hint 3: How do the magnitudes of h and c determine when quantum or relativistic effects will be observed?

Aufbau and Hund
(problem one hundred eight)

Hint 1: What are the ranges of the values for each quantum number?

Hint 2: What value of n corresponds to elements in the first period, second period, etc.? What value of l corresponds to s, p, d, and f orbitals?

Hint 3: How are spins aligned when multiple orbitals have identical energies?

Noble and Inert
(problem one hundred nine)

Hint 1: What is the electronic structure of the valence shell of a noble gas?

Hint 2: What property of fluorine and oxygen make them so reactive?

Hint 3: What is the ionization energy trend for the noble gas elements? What does this say about how strongly the outermost electrons are held, comparing, say He with Xe?

Heavy Metal
(problem one hundred ten)

Hint 1: What is the standard aufbau order of filling for orbitals?

Hint 2: How do half-filled orbitals provide an additional amount of stability to an atom? Can this alter the standard aufbau order of filling?

Hint 3: When cations are formed from transition metals, which are usually removed first, the s or d electrons?

Untrodden Paths
(problem one hundred eleven)

Hint 1: Chemical properties are related to the valence electron configuration. In which group would this element lie?

Hint 2: What is its electronic structure? Would it be relatively stable?

Hint 3: How would these structures make it compare in stability with other elements in its group? In its period?

Natural Attraction
(problem one hundred twelve)

Hint 1: What type of force is Mr. Phys describing? What other forces are there in the universe? How is the nucleus held together?

Hint 2: Are electrons truly particles that orbit the nucleus? What is the modern quantum mechanical conception of the atomic orbitals?

Hint 3: How can the wave-particle duality provide two alternative explanations for why electrons do not collapse into the nucleus?

Levels of Energy
(problem one hundred thirteen)

Hint 1: What does the energy of H depend on? Consider the simplicity of the atomic spectra of H as well as the variables in the Rydberg equation.

Hint 2: Why do the atomic spectra of helium and higher elements not have a simple series of bands as their spectra? Why do they not fit a simple Rydberg-like equation?

Hint 3: In a strong magnetic field, the atomic spectra of elements become even more complex. Hydrogen, however, remains the same. Which quantum numbers are responsible for describing the energy levels of orbitals in a magnetic field?

Orbitals and Orbits
(problem one hundred fourteen)

Hint 1: Are n=3 electrons necessarily further from the nucleus than n=2 or n=1 electrons according to Bohr's model? How about for the quantum mechanical model?

Hint 2: What is the relationship between the quantum numbers n and l and the number of nodes of the wave function?

Hint 3: Are s orbitals actually spherical shapes and p orbitals actually dumbells? If not, what do these theoretical surface boundaries signify?

Are You Affable?
(problem one hundred fifteen)

Hint 1: What is the definition of electron affinity? Does a negative electron affinity mean that the element can capture an electron readily?

Hint 2: Why do the halogens have a very negative electron affinity, whereas the alkali metals have a small electron affinity?

Hint 3: Where would an electron be placed if a noble gas atom captured an electron?

Identity Crisis (I)
(problem one hundred sixteen)

Hint 1: What is the trend of atomic radii across a period? Down a group?

Hint 2: How can the hint about having the largest atomic radius of the period pinpoint the group?

Hint 3: Which elements within this group have no d-electron shells filled?

Fashion Trends
(problem one hundred seventeen)

Hint 1: What part of their atomic structures is the same in the first set of ions/atoms? What is different?

Hint 2: What are the periodic trends for ionization energy and radius of atoms?

Hint 3: What are the electronic structures of Ca and K?

Identity Crisis (II)
(problem one hundred eighteen)

Hint 1: What are the trends of electron affinities and ionization energies across a period? Down a group?

Hint 2: What is (or might be) initially surprising about the EA and IE comparisons of elements N-1 and N+1? Consider your response to Hint 1, and if there is a problem try to resolve it.

Hint 3: Which elements are commonly found as ions but are toxic in their elemental forms?

Trinity
(problem one hundred nineteen)

Hint 1: What is the order of filling orbitals?

Hint 2: What are the trends for ionization energy and radius for elements with the same number of protons and different numbers of electrons?

Hint 3: What are the trends for ionization energy and radius for elements with the same number of electrons and different numbers of protons?

Identity Crisis (III)
(problem one hundred twenty)

Hint 1: What are the general atomic and ionic radii trends? Electronegativity trends?

Hint 2: Under what specific condition would the relative sizes of two atoms switch upon removal of an electron?

Hint 3: Which metals have great biological significance?

Chapter 8 — Bonding and Molecular Structure

Sharp Turns
(problem one hundred twenty-one)

What are the bond angles in a tetrahedral orbital configuration? How do the bond angles compare in CH_4, NH_3, and H_2O?

What are the bond angles in a trigonal planar orbital configuration? How do the bond angles compare in BF_3, NO_2, and NO_2^-?

What are the bond angles in a trigonal bipyramidal orbital configuration? How do the bond angles compare in PCl_5, SF_4, ClF_3, and I_3^-? Where are the lone pairs placed, in the equatorial or the axial orbitals? Why?

What are the bond angles in an octahedral orbital configuration? How do the bond angles compare in SF_6, BrF_5, and XeF_4?

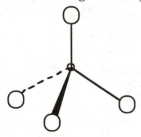

Tetrahedral

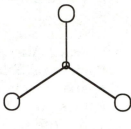

Trigonal Planar

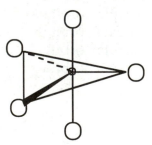

Trigonal Bipyramidal

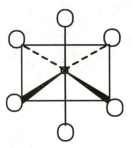

Octahedral

[category: bonding and molecular structure]
[topic: bonding]
[difficulty: A]

To Be or Not To Be?
(problem one hundred twenty-two)

To be or not to be polar, that is the question.

The compounds H_2N_2 and $[Co(NH_3)_4Cl_2]^+$ can each exist as two distinct species, one that is polar and one that is nonpolar. These isomeric species do not readily interconvert, and mixtures of the isomers can be physically separated. Draw the Lewis structures of the nonpolar and polar forms of H_2N_2. Describe their orbital geometries and draw valence-bond pictures of the two molecules. What makes one polar and the other nonpolar? Which one is the more stable (lower energy) species? Then answer the same questions for the complex ion $[Co(NH_3)_4Cl_2]^+$.

[category: bonding and molecular structure]
[topic: bonding]
[difficulty: A]

Platinum for Cancer
(problem one hundred twenty-three)

One of the major cancer chemotherapeutic agents is known as cis-platin. It is especially useful in testicular and ovarian cancers, and it functions by promoting cross-links in DNA.

The chemical formula for this compound is $Pt(NH_3)_2Cl_2$. It is also called cis-diamine-dichloro-platinum (II). The heavy metal atom (Pt) is the central atom, and the two chlorides and two ammonia molecules are situated around it. It is known that the geometric arrangement of these four substituents can alter dramatically the effectiveness of the drug. Is the molecular geometry tetrahedral or square planar? Draw the structures of two distinct molecular isomers, and circle which drug is active.

[category: bonding and molecular structure]
[topic: bonding]
[difficulty: A]

Life in a Line
(problem one hundred twenty-four)

A protein molecule is composed of amino acids that have the form NH_2-CHR-COOH (R is a variable side chain). With proper activation, they can polymerize to form a linear polypeptide that has the form -NH-CHR-CO-NH-CHR-CO-NH-etc. (see illustration). The CO-NH bond is called the peptide bond and is found to be rigid in character. It does not allow for free rotation around this bond. What resonance structure(s) must be significant in the polypeptide? Draw the relevant Lewis structures and show why this bond is rigid.

amino acid:

polypeptide:

peptide bond (rigid)

[category: bonding and molecular structure]
[topic: bonding]
[difficulty: A]

Sexy Figures
(problem one hundred twenty-five)

Two's company, but three's a crowd. With four, however, the numerous possibilities become very interesting. Describe the molecular structure of the following molecules or ions that have four fluorines: XeF_4, CF_4, SF_4, $[BF_4]^-$, $[ClF_4]^-$, $[ClF_4]^+$, $XeOF_4$, and SOF_4. Pay particular attention to the hybridization of the central atom, the geometry of the electronic orbitals, the positions of the lone pairs, and the overall structure. Which species are polar?

[category: bonding and molecular structure]
[topic: bonding]
[difficulty: B]

DNA Drama
(problem one hundred twenty-six)

The structure of DNA was first deduced by James Watson and Francis Crick (with the data and assistance of a number of other scientists) in 1952. It consists of two linear strands of repeating units called "nucleotides", and the two strands are hydrogen-bonded to each other and twisted around each other in space to form an overall double helix structure. DNA contains all the genetic information required for life within the nucleotide sequences. You will study this fascinating molecule in great detail if you take a course in molecular biology.

Each DNA nucleotide consists of a phosphate group, a sugar, and one of four "bases": adenine, thymine, guanine, or cytosine (see illustration). Thus there are four types of nucleotides in DNA. For each of the nucleotides (depicted below), describe its overall structure. Indicate which parts of the molecules are planar. What similarities among all the nucleotides do you notice?

Nucleotides:

R = sugar + phosphate

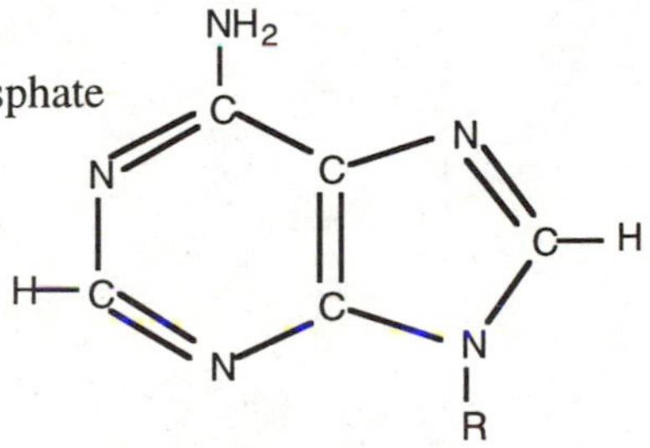

Adenine (A)

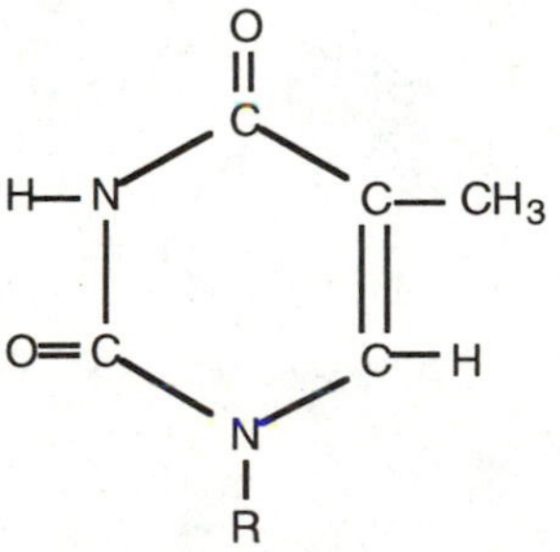

Thymine (T)

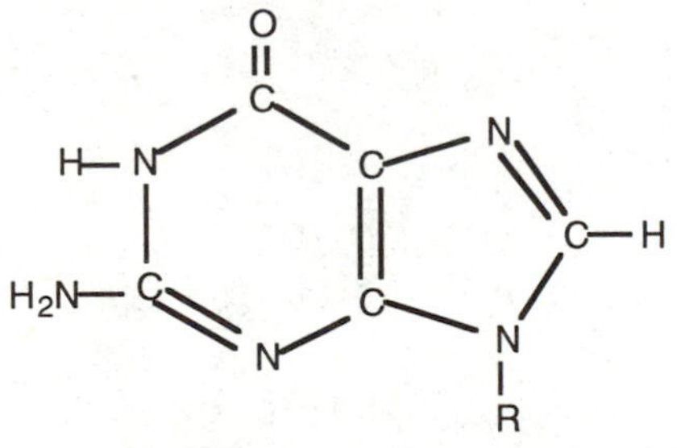

Guanine (G)

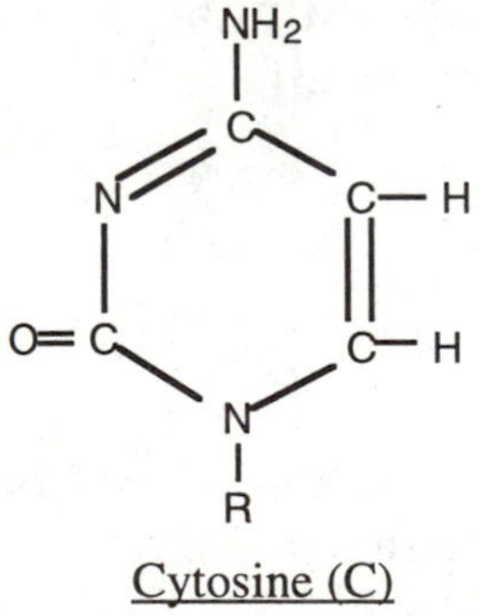

Cytosine (C)

[category: bonding and molecular structure]
[topic: bonding]
[difficulty: B]

Hewey Lewey
(problem one hundred twenty-seven)

You are doing your homework with a friend, and you two come up with different Lewis structures for H_4N_2 and H_4P_2. Your friend places a single bond between the central atoms and a lone pair of electrons on each. You, however, place a triple bond between the central atoms and include no lone pairs (see illustration). Both structures account for all the electrons and neither structure has any formal charge separation, yet only one is correct. Which is it and why? Is the other structure a *minor* species or a *nonexistent* species? Is there any difference between the nitrogen and phosphorus compounds?

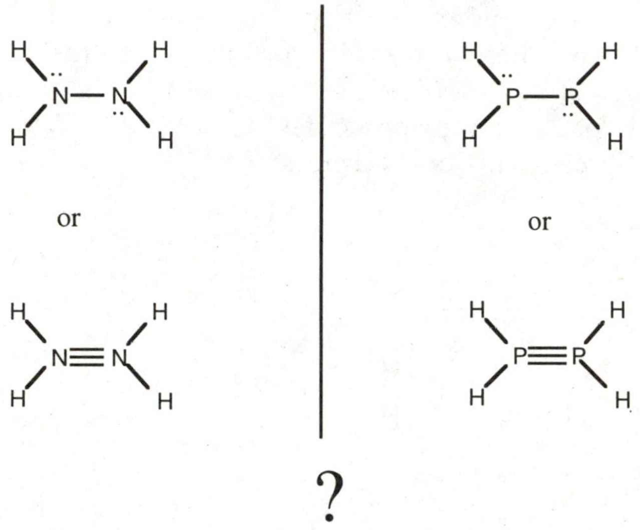

or

?

[category: bonding and molecular structure]
[topic: bonding]
[difficulty: B]

Planes and Solids
(problem one hundred twenty-eight)

PLANES: You have before you a set of compounds that all have ring structures with carbon backbones: C_6H_6, C_6H_{12}, $[C_5H_5]^-$, C_5H_5N, C_5H_6, and $C_{10}H_8$ (a double ring structure). In which of these structures are all the atoms in the same plane?

SOLIDS: What are possible three-dimensional, nonplanar structures for S_8, B_{12}, C_{60}, and P_4?

[category: bonding and molecular structure]
[topic: bonding]
[difficulty: B]

Spatial Relations
(problem one hundred twenty-nine)

Your chemistry professor draws a number of molecules on the blackboard: (1) CH_4; (2) $H_2C=CH_2$; (3) $H_2C=C=CH_2$; and (4) $H_2C=C=C=CH_2$. You muse about all the molecules that chemists draw on a two-dimensional blackboard and wonder which ones are actually planar, existing basically as they appear on the blackboard, and which ones are not planar, three-dimensional molecules. Which of the hydrocarbons above are planar? Describe the structure of the ones that are not planar, and provide a valence bond theory explanation for the trend.

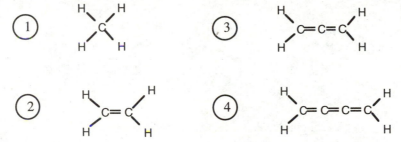

[category: bonding and molecular structure]
[topic: bonding]
[difficulty: C]

Planes and Convolutions
(problem one hundred thirty)

A very important molecule in red blood cells is known as heme (figure A). When red blood cells are degraded, the hemes are converted into a breakdown product called bilirubin. Although it can be envisioned as depicted on paper in figure B, certain indicated hydrogen bonds are formed so that it forms a compact, convoluted tertiary structure. Bilirubin is very hydrophobic and can be toxic. In normal physiology it is metabolized by the liver to a more soluble product and then excreted in stool. In pathological situations, bilirubin can build up, cause jaundice, and precipitate in the brain.

What is the overall molecular structure of a heme? What is the hybridization on all the atoms in the rings? Is there resonance in this structure? How would you describe the overall molecular structure of bilirubin? How do the hydrogen bonds enhance its hydrophobicity?

Figure A: A heme molecule:

Figure B: A bilirubin molecule (H-bonds indicated by arrows)

[category: bonding and molecular structure]
[topic: bonding]
[difficulty: C]

Too Close For Comfort
(problem one hundred thirty-one)

You are studying simple models of chemical bonding. In comparing the hydrogen molecule with the helium atom, you notice astutely that both have two protons surrounded by two electrons (see illustration).

Your teacher tells you that the hydrogen molecule has a characteristic bond length because of three factors: the attraction between the electrons and the nuclei, the repulsion between the two nuclei, and the repulsion between the two electrons. These factors lead to a minimum energy configuration (most stable structure) with a characteristic bond distance between the two nuclei. Sketch this energy versus distance plot. What occurs when the nuclei are too close? What occurs when the nuclei are further apart than their equilibrium bond length?

Wait a minute!!! How can this be true??? Surely helium is a stable structure, and yet the distance between the two protons is *much* less than in the hydrogen molecule!! What forces are responsible for holding the two protons so close together?

Why are there two seemingly stable configurations? What gives the helium nucleus its stability, and why does this not hold for the two nuclei of the hydrogen molecule?

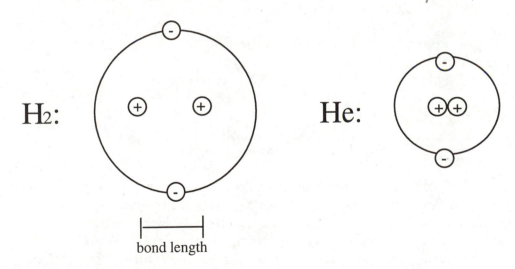

bond length

[category: bonding and molecular structure]
[topic: bonding]
[difficulty: C]

High on Helium
(problem one hundred thirty-two)

Describe the molecular orbital structure of diatomic helium, and state why it cannot exist under ordinary circumstances. Can it exist as a positive (+1) ion? Can it exist as a (-1) ion? Under what electronic conditions *can* it exist as a neutral species?

[category: bonding and molecular structure]
[topic: molecular orbitals]
[difficulty: A]

The Electric Co.
(problem one hundred thirty-three)

Electricity is promoted by electron mobility. The looser electrons are held to the nuclei, the more conductive a material is. Consider the following solids: diamond, graphite, and a slab of copper. What are the molecular structures and electron orbital structures of these solids? Are the electrons in bound states on individual nuclei or are they delocalized? Which ones conduct electricity the best and worst, based on their molecular structures?

[category: bonding and molecular structure]
[topic: molecular orbitals]
[difficulty: A]

Exciting Spins
(problem one hundred thirty-four)

In both atoms and molecules, electrons can be in the ground state or in an excited state. Consider an excited molecule in which one electron is in the π orbital and one electron is in the π* orbital. The excited molecule can undergo a chemical reaction, isomerize, or just return to the ground state. Returning to the ground state usually occurs by emitting a photon of light. This can occur by two pathways: by a fast mechanism (which takes about a nanosecond) causing fluorescence; or by a slow pathway (which takes seconds or longer) resulting in phosphorescence. The pathway that occurs depends on the spin states of the electrons. In the excited molecule, the electrons in the

π and π^* orbitals could have parallel or opposite spins (see illustration). Which spin configuration would result in fluorescence and which would result in phosphorescence? Why?

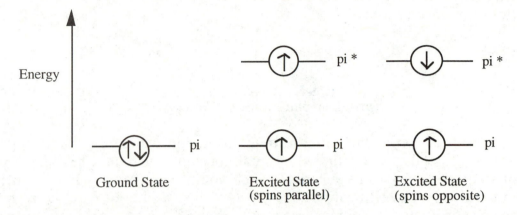

[category: bonding and molecular structure]
[topic: molecular orbitals]
[difficulty: B]

Beach Bums
(problem one hundred thirty-five)

You are lying on a Hawaiian beach, soaking up the sun's rays, and enjoying the warm tropical breeze tickling your face. <u>THIS</u> is how chemistry should be studied, you think. Your (kind) professor has paid for your trip in order for you to investigate the properties of air in an environment conducive to learning. To avoid disappointing him, you must prepare a report about the sand, the water, and the air. What are the two most important elements in each of these?

You decide to pick the air for further study. You separate the air into two main components, one that is paramagnetic and the other diamagnetic. You then expose each one to a strong electric field in order to form +1 ions. Would the bond length of the molecular ions be shorter or longer than the neutral molecules?   Identify each component and justify your answer with a molecular orbital argument.

You then go back to the beach for another afternoon in the sun. The major component of the air is relatively unreactive, you observe. (Fewer reactions to write about in your report!) Toward the end of the day, the sun becomes hidden behind the clouds, and a severe lighting storm erupts. You think that the major molecular species in air is now in an excited electronic state. How would the bond length in this excited (still neutral) molecule compare with the bond length in the ground state neutral molecule and the +1 cation?

[category: bonding and molecular structure]
[topic: molecular orbitals]
[difficulty: B]

Crazy Colors
(problem one hundred thirty-six)

Transition metals often form coordination complexes involving a central metal ion surrounded by small ions and/or molecules. These often are very colorful substances. Such complexes are very important in the study of metal-ligand chemistry as well as in the study of biology, since coordinated metal ions often play central roles in the functioning of complex biochemical systems.

According to crystal field theory, an octahedral arrangement of ligands would lead to the five d-orbitals splitting into two different energy levels— two orbitals have higher energy (e_g) and three orbitals have lower energy (t_{2g}). The difference between these energy levels is the crystal field splitting energy, and transitions between these levels are often in the visible range leading to color (see illustration). Use the above information and your knowledge of crystal field theory to describe why cuprous complexes are usually colorless but cupric complexes are usually colored. Remember, cuprous is Cu^+ and cupric is Cu^{2+}.

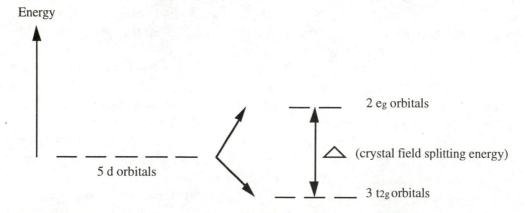

[category: bonding and molecular structure]
[topic: molecular orbitals]
[difficulty: B]

Long or Strong?
(problem one hundred thirty-seven)

Bond dissociation energies represent average energies of bonds over many molecules. You cannot merely ask what is the actual strength of a nitrogen-oxygen bond without specifying the molecule. Consider NO_2, N_2O, NO^+, NO_3^-, H_2NOH, and HNO. Order these six molecules in terms of increasing NO bond strength. Which one has the longest and which has the shortest NO bond?

[category: bonding and molecular structure]
[topic: molecular orbitals]
[difficulty: B]

Add and Subtract
(problem one hundred thirty-eight)

Molecular orbitals are formed from the combination of atomic orbitals. Orbital wave functions can add together in two manners. Just as waves in water or on a string can add to create constructive or destructive interference, atomic standing wave functions can either add positively or negatively in order to create molecular orbitals.

Consider first the molecular orbitals in the H_2 molecule. They are created from the combination of the 1s orbitals of the two H atoms. The atomic orbitals can be combined constructively to form the $\Psi_{1s\,(atom\,1)} + \Psi_{1s\,(atom\,2)}$ orbital or destructively to form the $\Psi_{1s\,(atom\,1)} - \Psi_{1s\,(atom\,2)}$ orbital. Which one is the σ_{1s} bonding molecular orbital? Which one is the σ^*_{1s} antibonding molecular orbital? How do these expressions suggest shapes for the molecular orbitals? Note that one has increased electron density between the nuclei, and the other one has decreased electron density between the nuclei.

Consider now the molecular orbitals that can be formed from the combination of two 2p orbitals. What are the four ways that they can combine (mathematically and geometrically)? Think about both constructive and destructive combinations as well as their relative geometrical orientations. What are the resulting shapes of the molecular orbitals? These form the σ_{2p}, π_{2p}, π^*_{2p}, and σ^*_{2p} orbitals. Why is there one σ_{2p} and one σ^*_{2p} orbital but two π_{2p} and two π^*_{2p} orbitals?

[category: bonding and molecular structure]
[topic: molecular orbitals]
[difficulty: B]

NO!!
(problem one hundred thirty-nine)

You are a researcher investigating the magnetic properties of nitrogen and oxygen. Recall the Stern-Gerlach experiment in which a beam of silver atoms (each with one unpaired electron) was subjected to an inhomogeneous magnetic field, which separated the atoms into two beams — one representing atoms containing the unpaired electron with a +1/2 spin and the other representing atoms containing the unpaired electron with the opposite -1/2 spin. The magnetic field thus has the ability to separate particles on the basis of their spin states.

You perform similar experiments with beams of either molecular nitrogen or oxygen. Into how many beams would each one be split? At low temperatures, which beam(s) and spin conformations would be strongest? At high temperatures, which beam(s) and spin conformations would be the strongest? Why does temperature matter?

(1) Stern-Gerlach:

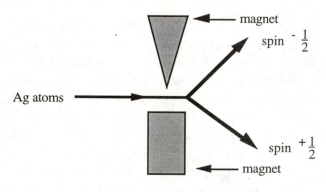

strong magnetic field

(2) Molecular nitrogen or oxygen:

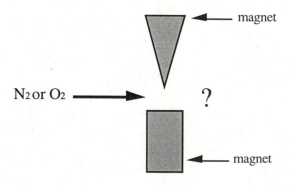

strong magnetic field

[category: bonding and molecular structure]
[topic: molecular orbitals]
[difficulty: C]

Ho, Ho, Ho
(problem one hundred forty)

Consider two species: the hydroxyl radical (HO) and the hydroxide ion (HO⁻). For each, construct an energy correlation diagram for their molecular orbitals (bonding, nonbonding, and antibonding). You can use the template shown in the illustration. Note that the 1s orbital of the H has a higher energy than the 1s, 2s, or even the 2p orbitals of O. What is the bond order of the H-O bond in each case? Does the bonding σ orbital contain more H character, more O character, or an equal mixture? Does the antibonding σ* orbital contain more H character, more O character, or an equal mixture?

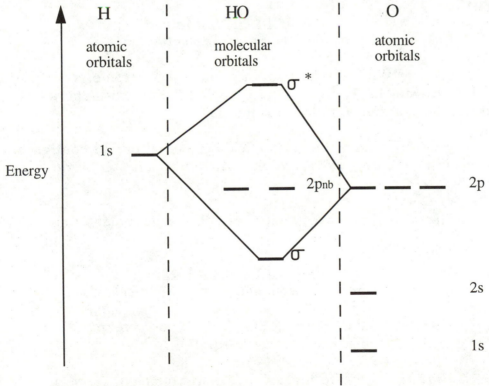

[category: bonding and molecular structure]
[topic: molecular orbitals]
[difficulty: C]

Bonding and Molecular Structure — HINTS

Sharp Turns
(problem one hundred twenty-one)

Hint 1: What are the hybridizations of the central atom in each case?

Hint 2: What are the Lewis structures? How many lone pairs does each have?

Hint 3: Geometrically, which is the "largest", the size of a bonding orbital, an orbital with a lone pair, or the an orbital with a single electron?

To Be or Not To Be?
(problem one hundred twenty-two)

Hint 1: What are the hybridization and electronic geometries of each nitrogen in H_2N_2? What type of bond exists between the nitrogens? Is it free to rotate or is it rigid?

Hint 2: What is the crystal field octahedral orbital configuration of $[Co(NH_3)_4Cl_2]^+$? How are the ammonia and chloride ligands attached?

Hint 3: What makes molecules polar or nonpolar?

Platinum for Cancer
(problem one hundred twenty-three)

Hint 1: In the formula PtX_2Y_2 can you have more than one configuration if the structure is tetrahedral?

Hint 2: In the formula PtX_2Y_2 can you have more than one configuration if the structure is square planar?

Hint 3: What does "cis" mean?

Life in a Line
(problem one hundred twenty-four)

Hint 1: Is the bond in CH_3-CH_3 rigid? How about CH_2=CH_2?

Hint 2: What must be the hybridization of the nitrogen in the peptide bond?

Hint 3: Are the two possible structures equally favored, or is one more stable than the other?

Sexy Figures
(problem one hundred twenty-five)

Hint 1: How many valence electrons does the central atom of each molecule have? How can this help you draw the Lewis structure of each?

Hint 2: From the Lewis structure, determine the hybridization of each central atom. How can VSEPR (valence shell electron pair repulsion theory) be used to orient the orbitals in the proper conformations? How can VSEPR also be used to place the lone pairs appropriately?

Hint 3: From the structures, determine which bonds have dipole moments. Does the (a)symmetry of the molecule determine whether the dipoles cancel each other out?

DNA Drama
(problem one hundred twenty-six)

Hint 1: In the ring structures, what are the hybridizations of the atoms that are part of the ring?

Hint 2: How can you use VSEPR theory to determine which atoms have a "trigonal planar" arrangement of orbitals? How about "tetrahedral" or other orbital arrangements?

Hint 3: How can you use the geometry of each atom of a ring to deduce the overall structure of the entire ring?

Hewey Lewey
(problem one hundred twenty-seven)

Hint 1: What are the possible Lewis structures described for each?

Hint 2: Which orbitals are available for bonding in nitrogen and phosphorus? Are they the same?

Hint 3: Are the differences between the structures relatively minor in importance? Or are there fundamental reasons why a structure could never exist? It might help to make a list of the advantages/disadvantages of each.

Planes and Solids
(problem one hundred twenty-eight)

Hint 1: What are the Lewis structures of each compound?

Hint 2: What are the hybridization of the atoms that comprise the ring?

Hint 3: How many vertices are there in an icosahedron? After which other geometrical shapes might complex molecules be modeled?

Spatial Relations
(problem one hundred twenty-nine)

Hint 1: What is the hybridization of the carbon in CH_4? The carbons in $H_2C=CH_2$? The central carbons in the other molecules?

Hint 2: What type of bonds (sigma, pi, delta, etc.) are present in each molecule? From which orbitals on each atom are each of the bonds made?

Hint 3: What is the <u>relative orientation</u> of the orbitals that make each bond?

Planes and Convolutions
(problem one hundred thirty)

Hint 1: From VSEPR theory, what is the structure of a central atom with three groups around it?

Hint 2: In a heme, would any of the bonds allow free rotation? Or is the structure rigid? Which bonds in bilirubin allow free rotation?

Hint 3: How does hydrophobicity or hydrophilicity relate to the hydrogen bonding potential of a molecule in aqueous solution?

Too Close For Comfort
(problem one hundred thirty-one)

Hint 1: How similar are the particles that comprise a hydrogen molecule and helium atom?

Hint 2: What are the effective distances and relative strengths of the forces involved in each case?

Hint 3: Why is the H_2 <—> He interconversion not observed?

High on Helium
(problem one hundred thirty-two)

Hint 1: From which orbital will an electron be removed to make a positive ion?

Hint 2: In which orbital could an electron be placed to make a negative ion? Is this realistic?

Hint 3: What electronic configurations are necessary to make neutral diatomic helium?

The Electric Co.
(problem one hundred thirty-three)

Hint 1: What is the Lewis structure for diamond (C), which is a covalently bonded complex with tetrahedral lattice points?

Hint 2: What is the Lewis structure for graphite (C), which has only trigonal planar lattice points?

Hint 3: What are the shapes and properties of the electron clouds a metal?

Exciting Spins
(problem one hundred thirty-four)

Hint 1: What is Hund's rule? What is the Pauli principle?

Hint 2: Which spin conformation would lead to more rapid photon emission?

Hint 3: Do spins change during normal photon emission?

Beach Bums
(problem one hundred thirty-five)

Hint 1: What are the two main (diatomic) gases in the air?

Hint 2: Write down the molecular orbital descriptions of each. Which one is paramagnetic? Is the highest energy electron in each in a bonding or antibonding orbital?

Hint 3: In forming an ion or excited atom, how does the electronic change affect bond order and bond length?

Crazy Colors
(problem one hundred thirty-six)

Hint 1: How many d-electrons are in the Cu^+ and Cu^{2+} ions?

Hint 2: How are the valence electrons distributed between the e_g and t_{2g} orbitals?

Hint 3: Can transitions between these energy levels occur?

Long or Strong?
(problem one hundred thirty-seven)

Hint 1: What are the Lewis structures of all these species?

Hint 2: Which have resonance? What are the bond orders of each?

Hint 3: Does a higher bond order mean a longer or shorter bond? Stronger or weaker?

Add and Subtract
(problem one hundred thirty-eight)

Hint 1: Is a bonding orbital one that has increased or decreased electron density in the bond?

Hint 2: Two spherically symmetric s orbitals can exist in only one relative geometry. In which two fundamental geometries can the axially symmetric p orbitals exist? Which geometry gives rise to σ and π bonds?

Hint 3: Would addition or subtraction of the 2p atomic wave functions yield the sigma constructive interference (creating the σ_{2p} bonding orbital)? Would addition or subtraction of the 2p atomic wave functions yield the pi constructive interference (creating the π_{2p} bonding orbital)?

NO!!
(problem one hundred thirty-nine)

Hint 1: What determines whether an atom/molecule is paramagnetic or diamagnetic?

Hint 2: What is Hund's Rule? How does this affect which spin conformation is most common?

Hint 3: High temperature imparts more overall energy to each molecule, thus making spin-spin pairing energies less important. Does this imply that Hund's Rule is more accurate at low or high temperatures?

Ho, Ho, Ho
(problem one hundred forty)

Hint 1: Does the 1s and 2s orbitals of O mix with the 1s orbital of H?

Hint 2: Does the 2p orbitals of O mix with the 1s orbital of H? What molecular orbitals do these four atomic orbitals "become"?

Hint 3: How do the relative energies of the valence atomic orbital determine whether the bonding orbital will have more H or more O character?

Chapter 9 — Special Topics

Ozone Hole
(problem one hundred forty-one)

In the stratosphere, photochemistry plays a crucial role in absorbing ultraviolet light. As discussed in problem 101 "Fun in the Sun", UV light is very dangerous to living species, and thus most of it must be prevented from penetrating the atmosphere.

(i) Why is UV light more dangerous to life than visible light? Ozone absorbs much of the "harmful" UV radiation (200-350 nm) in a photodissociation as follows:

$$O_3 + h\nu \longrightarrow O_2 + O$$

(ii) Chlorofluorocarbons (CFCs) such as CCl_2F_2 are thought to deplete the ozone in the atmosphere by forming reactive chlorine atoms. Describe in a qualitative way how absorption of a photon can lead to photodissociation of a bond.

$$CCl_2F_2 + h\nu \longrightarrow CClF_2 + Cl$$

(iii) In general, why is atomic chlorine so reactive? One of the reactions of chlorine is to catalyze the following ozone destruction reaction (which actually occurs by a complex multi-step mechanism):

$$2O_3 \longrightarrow 3O_2$$

(iv) Atmospheric ozone can be replaced (through a series of other photochemical reactions), but if the destruction exceeds new production, then there will be a net depletion of atmospheric ozone. This has been particularly true above the Antarctic but also is a global problem. What do you think would be the result of continued large-scale production of CFCs to life on earth?

[category: special topics]
[topic: photochemistry]
[difficulty: A]

Can You See It?
(problem one hundred forty-two)

Upon absorbing a photon of light, a molecule can become excited and then has a number of fates. It could emit the energy by releasing a photon; it could undergo photodissociation, in which a bond breaks and a chemical reaction occurs; it could react with another molecule; or it could undergo photoisomerization, in which the the excited molecule rearranges itself and achieves a "different" ground state.

An example of photoisomerization involves a molecule called retinal. It is a pigment in your eye that isomerizes when it absorbs a photon of light. This is the primary event in vision. The isomerized retinal then sets off a cascade of complex biochemical and neurological reactions that result in vision. Depicted in the illustration is the isomerization of retinal. How would you describe the isomerization (look specifically at the bond indicated by the arrow). Why does this change not occur without photoexcitation? How would an excited state of retinal facilitate this isomerization?

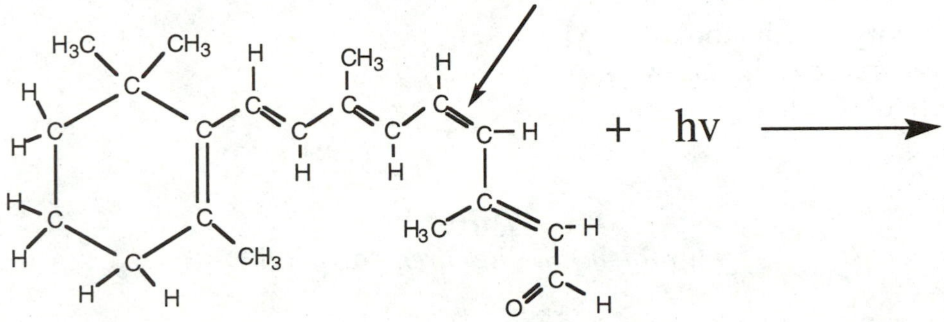

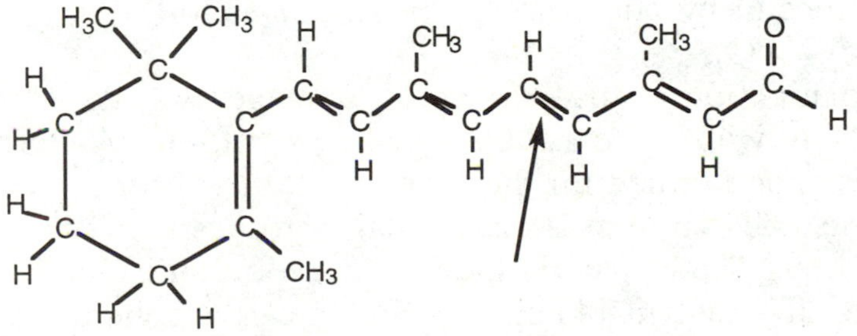

[category: special topics]
[topic: photochemistry]
[difficulty: B]

Clean and Hot
(problem one hundred forty-three)

Here are two common examples of chemical radicals:

One involves the dry cleaning industry. CCl_4 is a chemical once used in dry cleaning, and it is metabolized by an enzyme in the liver (called P450) as follows: CCl_4 combines with an electron to form CCl_3 and a chloride ion. Write out this reaction, clearly indicating which substances are radicals and which are not. Show that there are equal numbers of electrons on both sides of the reaction (ie, that the reaction is balanced). What are the molecular structures of CCl_4 and CCl_3?

One of the dangers of exposure to radiation is the generation of reactive radicals. High energy radiation is known to cause the photodissociation of water to form atomic hydrogen and hydroxyl radicals. Write out this reaction, clearly indicating which substances are radicals and which are not. Show that there are equal numbers of electrons on both sides of the reaction (i.e., that it is balanced).

[category: special topics]
[topic: radical chemistry]
[difficulty: A]

Radical!
(problem one hundred forty-four)

A number of chemical radicals are important to normal physiological processes as well as pathological processes in the human body. The superoxide ion can be produced from oxygen molecules by either oxidase enzymes or by nonenzymatic methods as follows:

$$O_2 + e^- \longrightarrow O_2^-$$

This superoxide ion causes severe damage to the lipid membranes of cells as well as to cellular proteins and DNA. Such damage can either be good or bad for the individual: it is "good" when it is used by immune cells known as macrophages to destroy *invading* bacteria; however it is "bad" when it damages your *own* cells and DNA. Such is often the case in biology — that a certain chemical or physiological process is beneficial when it occurs in a controlled fashion but is harmful when it is uncontrolled. In many circumstances the unwanted superoxide ions can be degraded by an enzyme called superoxide dismutase before they cause any harm.

Consider the oxygen molecule and superoxide ion. What are the molecular orbital descriptions of each? What are their bond orders? Which one is more reactive and why? One of the damaging reactions is the combination of the superoxide ion with nitric oxide to form a hydroxide radical and nitrogen dioxide. Write out this reaction, and draw Lewis structures of all the molecules.

[category: special topics]
[topic: radical chemistry]
[difficulty: B]

Transformation
(problem one hundred forty-five)

For each of the following, state whether the element's identity will change and if so, by how much its atomic number will increase or decrease:
- release of an alpha particle;
- release of a negative beta particle;
- release of a gamma ray;
- nuclear fission;
- loss of an electron (ionization); and
- change in energy level of an electron.

[category: special topics]
[topic: nuclear chemistry]
[difficulty: A]

Funny Fluorescence
(problem one hundred forty-six)

The French scientist Becquerel was studying various fluorescent materials. He took materials, irradiated them by ultraviolet light to excite electrons, and placed them on X-ray film which recorded the emission of photons. When he took a piece of uranium ore, he found that it made a print on X-ray film even when it had not been irradiated with light. Why is this surprising, and what is the explanation for this phenomenon?

[category: special topics]
[topic: nuclear chemistry]
[difficulty: B]

Wish Upon a Star
(problem one hundred forty-seven)

You start as H inside a star— Now I wonder what you are—
One of the major reactions that fuels the sun's energy is the fusion of four hydrogen atoms to produce a helium atom and two positrons. (Note that the actual process involves a more complicated

multi-step reaction.) Fusion of helium, then of carbon, etc. is then thought to form the heavier elements. Write a balanced nuclear reaction for the formation of He from H. Is mass and charge conserved in this process? Given that this reaction releases energy, which would you expect to be more massive, the reactants or the products?

Why does a very small mass deficit cause such a huge amount of energy to be released?

[category: special topics]
[topic: nuclear chemistry]
[difficulty: B]

Rope Trick
(problem one hundred forty-eight)

Nylon is a synthetic polymer that can be formed at the interface of two immiscible liquids. It can be drawn out from this interface as a continuous rope (see illustration). The two reagents you need are hexamethylenediamine, $H_2N(CH_2)_6NH_2$, which is dissolved in an aqueous solution and sebacoyl chloride, $ClCO(CH_2)_8COCl$, which is dissolved in hexane (an organic solvent). The first reagent is called a diamine, due to the presence of the two -NH_2 groups, and the second is called an diacid chloride, due to the presence of the two -COCl groups. Noting the symmetry of the molecules, write out the interfacial condensation reaction, that is, the polymerization reaction that forms a seemingly endless thread of nylon. For best results, should you make the aqueous phase of the solution acidic or basic?

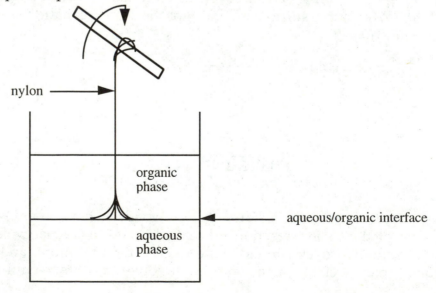

[category: special topics]
[topic: misc. topics]
[difficulty: A]

Slicing Water
(problem one hundred forty-nine)

Corn starch is a common culinary product that is used for thickening sauces, soups, and other liquids. It comes as a white powder and can easily dissolve in a water-based solution to increase its viscosity. If you make a saturated solution of corn starch in water, then you have a soupy white substance that is very viscous. This is called a "thixotropic solution". If you fill a glass with this mixture and plunge a knife into it very slowly, then you can, with relative ease, touch the bottom of the glass. If, however, you attempt to plunge the knife into the glass quickly, then the solution solidifies transiently and prevents you from hitting the bottom of the glass! When you remove the knife, then the solution liquefies again! Try this yourself next time you are in a kitchen!

Note that starch is a long, heavily branched polymer of sugar residues. What might be happening in both cases above?

[category: special topics]
[topic: misc. topics]
[difficulty: B]

Mythology and Folklore
(problem one hundred fifty)

Many chemistry courses and textbooks have claimed that the following statements are true, yet they are not. They are all incorrect (at least in part). State why the following are wrong, give at least one counterexample for each, and write a corrected version of each of these statements.

(i) All atoms in Lewis structures should have an octet of electrons if the molecule is to be stable.

(ii) The scale of pH is from 0 to 14.

(iii) The conjugate of a strong acid or base is weak, and the conjugate of a weak acid or base is strong.

(iv) An acid and a base react to yield a salt plus water.

(v) If there were no man-made acid rain effects, then the water you would be drinking would have a pH of 7.

(vi) The noble gases are unreactive.

(vii) A bond is formed by two atoms each donating one electron and both sharing the electrons equally.

(viii) Bonds fall into two distinct categories: ionic and covalent.

(ix) An atom can be thought of as a solar system, with the nucleus in the center and orbiting electrons.

(x) Chemistry is difficult.

[category: special topics]
[topic: misc. topics]
[difficulty: B]

Special Topics — HINTS

Ozone Hole
(problem one hundred forty-one)

Hint 1: What determines the energy of electromagnetic radiation?

Hint 2: What does addition of energy do to a ground state molecule?

Hint 3: What is the electronic structure and electronegativity of atomic chlorine?

Can You See It?
(problem one hundred forty-two)

Hint 1: Are there any atomic differences in the two states of retinal? How about structural differences?

Hint 2: What terminology would you use to describe the double bond indicated in each structure?

Hint 3: Does the double bond allow free rotation in the ground state? In the excited state?

Clean and Hot
(problem one hundred forty-three)

Hint 1: Does a radical fulfill the "octet rule"?

Hint 2: In determining molecular structure, how do you include a single electron in determining the overall "steric number" or "electronic geometry"?

Hint 3: Must the number of electrons be balanced in a reaction?

Radical!
(problem one hundred forty-four)

Hint 1: What makes a radical so reactive?

Hint 2: If you add an electron to molecular oxygen, do you add a bonding or antibonding electron?

Hint 3: How is the Lewis structure of superoxide different from oxygen, and how does this reflect the bond orders?

Transformation
(problem one hundred forty-five)

Hint 1: What is the difference between a nuclear reaction and an electron reaction?

Hint 2: What are alpha, beta, and gamma particles?

Hint 3: What quantities are conserved in a chemical reaction? In a nuclear reaction? What determines an element's identity?

Funny Fluorescence
(problem one hundred forty-six)

Hint 1: What is the phenomenon of fluorescence? Why is irradiation necessary?

Hint 2: Is the uranium exhibiting fluorescence? If so, why does it not need to be irradiated? If not, what other possible processes might it be undergoing?

Hint 3: Becquerel's work came to a rather surprising conclusion that opened up a new branch of chemistry and helped advance the current notions of atomic structure. What's the difference between the reaction of a fluorescent material and the reaction of the uranium ore?

Wish Upon a Star
(problem one hundred forty-seven)

Hint 1: What quantities are always conserved in a nuclear reaction? Is this conservation approximate or absolute?

Hint 2: What subatomic reaction can occur between protons, neutrons, and electrons? This is essential for a number of processes including this fusion reaction as well as beta decay.

Hint 3: What's the relation between mass deficit and energy released?

Rope Trick
(problem one hundred forty-eight)

Hint 1: What are the Lewis structures of the two reagents?

Hint 2: Why can you only get polymerization at the aqueous-organic interface? Why then can the nylon rope be much longer than the diameter of the beaker containing the two reagents?

Hint 3: Is the hexamethylenediamene a weak acid or base? What would its structure be in a low and a high pH environment? What pH would be necessary for it being active in a polymerization reaction?

Slicing Water
(problem one hundred forty-nine)

Hint 1: What is the definition of viscosity? How is the retarding force of an object moving through a viscous medium affected by its velocity? Can the observation be accounted for solely on the basis of viscosity?

Hint 2: What is the structure of sugar and starch molecules? What reactions between them can occur in a very concentrated solution?

Hint 3: What variable changes when the knife is plunged in quickly? Can this affect one of these reactions?

Mythology and Folklore
(problem one hundred fifty)

Hint 1: Which are the only elements that usually follow the "octet rule"?

Hint 2: What is the definition of pH? Can it exceed these boundaries?

Hint 3: How are "strong" and "weak" defined? Is a conjugate pair always strong/weak or weak/strong? What are the reactions of non-Arrhenius acids and bases?